SpringerBriefs in Mathematics

SpringerBriefs in Mathematics showcases expositions in all areas of mathematics and applied mathematics. Manuscripts presenting new results or a single new result in a classical field, new field, or an emerging topic, applications, or bridges between new results and already published works, are encouraged. The series is intended for mathematicians and applied mathematicians.

BCAM SpringerBriefs

BCAM *SpringerBriefs* aims to publish contributions in the following disciplines: Applied Mathematics, Finance, Statistics and Computer Science. BCAM has appointed an Editorial Board, who evaluate and review proposals.

Typical topics include: a timely report of state-of-the-art analytical techniques, bridge between new research results published in journal articles and a contextual literature review, a snapshot of a hot or emerging topic, a presentation of core concepts that students must understand in order to make independent contributions.

Please submit your proposal to the Editorial Board or to Francesca Bonadei, Executive Editor Mathematics, Statistics, and Engineering: francesca.bonadei@springer.com.

More information about this series at http://www.springer.com/series/10030

Xiaoyu Fu · Qi Lü · Xu Zhang

Carleman Estimates for Second Order Partial Differential Operators and Applications

A Unified Approach

Xiaoyu Fu
School of Mathematics
Sichuan University
Chengdu, China

Qi Lü
School of Mathematics
Sichuan University
Chengdu, China

Xu Zhang
School of Mathematics
Sichuan University
Chengdu, China

ISSN 2191-8198   ISSN 2191-8201   (electronic)
SpringerBriefs in Mathematics
ISBN 978-3-030-29529-5   ISBN 978-3-030-29530-1   (eBook)
https://doi.org/10.1007/978-3-030-29530-1

This Springer imprint is published by the registered company Springer Nature Switzerland AG
The registered company address is: Gewerbestrasse 11, 6330 Cham, Switzerland

Preface

In a seminal paper [3], T. Carleman introduced a celebrated method to prove the strong unique continuation property for second order elliptic partial differential equations (PDEs) with two variables. His method, now known as the Carleman estimate, has become one of the major tools in the study of unique continuation properties (e.g., [8, 11]), uniqueness and stability of Cauchy problems (e.g., [2, 9, 23]), inverse problems (e.g., [1, 4, 10, 12, 14, 18]), and control of both deterministic and stochastic PDEs (e.g. [5–7, 13, 15–17, 19–22]). Results on this topic are distributed in many papers/monographs. On the other hand, generally speaking, Carleman estimates depend strongly on the type and nature of the underlying equations. Hence, it is not convenient for many beginners to learn these techniques systematically.

The Carleman estimate is actually quite elementary. It is simply a weighted energy estimate. A rudiment of this method is even available in the setting of ordinary differential equations, say the classical integrating factor method. Anyone who knows elementary calculus can grasp the main idea of the Carleman estimate in a few minutes [see Proposition 1.1 in Chap. 1 for a stability estimate for linear ordinary differential equations (ODEs), for which we present a Carleman estimate-based proof; see also Remark 1.1 for a little more explanation].

In this book, we give a brief and (almost) self-contained introduction to Carleman estimates for three typical PDEs of second order, i.e., elliptic, parabolic, and hyperbolic equations, and their typical applications in control, unique continuation, and inverse problems (though we focus a little more on control problems). The main novelties of this book are as follows:

- Only some basic calculus is needed to obtain the main results in this book, though some elementary knowledge of functional analysis and PDEs will be helpful in understanding them.
- All the Carleman estimates in the book are derived from a fundamental identity for second order partial differential operators (PDOs). The only difference is the choice of weight functions.

- Only some rather weak smoothness and/or integrability conditions are needed for the coefficients appearing in the equations.

Due to space limitations, many interesting Carleman estimates for second order PDEs (and their applications) are not covered here, such as the L^p Carleman estimate, Carleman estimate with a limiting weight function, Carleman estimate for equations with jump coefficients, Carleman estimate for coupled PDEs, and Carleman estimate for stochastic PDEs. As a remedy, in the text we give the corresponding references.

The authors would like to thank Prof. Enrique Zuazua for his kind invitation to write this book and Dr. Zhongqi Yin, who read the book so carefully and gave us many useful suggestions/comments. This work is partially supported by NSF of China under grants 11221101, 11931011, 11971333 and 11971334 the NSFC-CNRS Joint Research Project under grant 11711530142, the PCSIRT under grant IRT 16R53, and the Chang Jiang Scholars Program from the Chinese Education Ministry.

Chengdu, China   Xiaoyu Fu
October 2019   Qi Lü
                                                                                             Xu Zhang

References

1. Bellassoued, M., Yamamoto, M.: Carleman Estimates and Applications to Inverse Problems for Hyperbolic Systems. Springer Monographs in Mathematics. Springer Japan KK (2017)
2. Calderón, A. P.: Uniqueness in the Cauchy problem for partial differential equations. Amer. J. Math. **80**, 16–36 (1958)
3. Carleman, T.: Sur un problème d'unicité pour les systèmes d'équations aux dérivées partielles à deux variables indépendantes. Ark. Mat. Astr. Fys. 26 B. **17**, 1–9 (1939)
4. Choulli, M.: Applications of Elliptic Carleman Inequalities to Cauchy and Inverse Problems. Springer Briefs in Mathematics. BCAM Springer Briefs. Springer, Cham (2016)
5. Fernández-Cara, E., Guerrero, S., Imanuvilov, O. Yu., Puel, J.-P.: Local exact controllability of the Navier-Stokes system. J. Math. Pures Appl. **83**, 1501–1542 (2004)
6. Fernández-Cara, E., Zuazua, E.: Null and approximate controllability for weakly blowing up semilinear heat equations. Ann. Inst. H. Poincaré Anal. Non Linéaire. **17**, 583–616 (2000)
7. Fursikov, A.V., Imanuvilov, O. Yu.: Controllability of Evolution Equations. Lecture Notes Series 34, Seoul National University, Seoul, Korea (1996)
8. Hörmander, L.: The Analysis of Linear Partial Differential Operators. III. Pseudodifferential Operators. Springer, Berlin (1985)
9. Hörmander, L.: The Analysis of Linear Partial Differential Operators. IV. Fourier Integral Operators. Springer, Berlin (1985)
10. Isakov, V.: Inverse Problems for Partial Differential Equations. Springer, New York (2006)
11. Kenig, C.E.: Carleman estimates, uniform Sobolev inequalities for second order differential operators, and unique continuation theorems. International Congress of Mathematicians, Vol. II, Berkeley, California, USA, pp. 948–960 (1986)
12. Klibanov, M.V.: Carleman estimates for global uniqueness, stability and numerical methods for coefficient inverse problems. J. Inverse Ill-Posed Probl. **21**, 477–560 (2013)

13. Lebeau, G., Le Rousseau, J., Robbiano, L.: Elliptic Carleman Estimates and Applications to Stabilization and Controllability. Volume 1: Dirichlet boundary condition on Euclidean space. Book in preparation.
14. Lü, Q.: Carleman estimate for stochastic parabolic equations and inverse stochastic parabolic problems. Inverse Problems, **28**, 045008 (2012)
15. Lü, Q.: Observability estimate for stochastic Schrödinger equations and its applications. SIAM J. Control Optim. **51**, 121–144 (2013)
16. Lü, Q.: Exact controllability for stochastic Schrödinger equations. J. Differential Equations **255**, 2484–2504 (2013)
17. Lü, Q.: Exact controllability for stochastic transport equations. SIAM J. Control Optim. **52**, 397–419 (2014)
18. Yamamoto, M.: Carleman estimates for parabolic equations and applications. Inverse Problems. **25**, 123013 (2009)
19. Zhang, X.: Exact Controllability of the Semilinear Distributed Parameter System and Some Related Problems. Ph.D. Thesis, Fudan University, Shanghai, China (1998)
20. Zhang, X.: Explicit observability inequalities for the wave equation with lower order terms by means of Carleman inequalities. SIAM J. Control Optim. **39**, 812–834 (2001)
21. Zhang, X.: A unified controllability/observability theory for some stochastic and deterministic partial differential equations. In: Proceedings of the International Congress of Mathematicians, Vol. IV, Hyderabad, India, pp. 3008–3034 (2010)
22. Zuazua, E.: Controllability and observability of partial differential equations: some results and open problems. In Handbook of Differential Equations: Evolutionary Differential Equations, vol. 3, Elsevier Science, pp. 527–621 (2006)
23. Zuily, C.: Uniqueness and Non-Uniqueness in the Cauchy Problem. Birkhäuser, Boston (1983)

Contents

About the Authors

Xiaoyu Fu is Professor of Mathematics in the School of Mathematics, Sichuan University, Chengdu, China. Her main research interest is control theory of partial differential equations.

Qi Lü is a Professor in the School of Mathematics, Sichuan University, Chengdu, China. His main research interest is Mathematical Control Theory, including in particular control theory of deterministic and stochastic partial differential equations.

Xu Zhang is Cheung Kong Scholar Distinguished Professor in the School of Mathematics, Sichuan University, Chengdu, China. His main research interests include mathematical control theory and related partial differential equations and stochastic analysis.

Chapter 1
Introduction

Abstract In this chapter, we shall establish a fundamental weighted identity for second order partial differential operators, via which the main results (Carleman estimates and applications) in this book and most of other related results in previous references can be deduced. Also, some frequently used notations (throughout this book) will be introduced, and some background for Carleman estimates and two stimulating examples explaining the main idea of these sort of estimates will be presented.

Keywords Weighted identity · Second order partial differential operator · Carleman estimate

1.1 Some Notations and Background

Throughout this book, $\mathbb{N}$, $\mathbb{R}$ and $\mathbb{C}$ stand for respectively the sets of positive integers, real numbers and complex numbers. For a complex number c, we denote by $\bar{c}$, Re c and Im c, its complex conjugate, real part and imaginary part, respectively. Write $i = \sqrt{-1}$.

In what follows, we fix $T > 0$, $m, n \in \mathbb{N}$ and a bounded domain Ω of $\mathbb{R}^n$ with a C^2 boundary $\Gamma \equiv \partial\Omega$. Let Γ_0 be a nonempty open subset of Γ, and let ω_0, ω be nonempty open subsets of Ω satisfying $\overline{\omega}_0 \subset \omega$. Put

$$Q \overset{\triangle}{=} (0, T) \times \Omega, \qquad \Sigma \overset{\triangle}{=} (0, T) \times \Gamma.$$

Denote by χ_ω the characteristic function of ω, and by $\nu(= (\nu^1, \nu^2, \ldots, \nu^n)) = \nu(x)$ the unit outward normal vector of Ω at $x \in \Gamma$. We shall use some standard notations from Sobolev spaces (e.g. [1]) and semigroup theory (e.g. [59]). Sometimes, we use the notion $u \in H^1(\Omega; \mathbb{C})$ or $u \in H^1(\Omega; \mathbb{R})$ to emphasize that u is complex-valued or real-valued function (If there is no need to emphasize this, we simply use the notation $H^1(\Omega)$). Nevertheless, we shall simply denote its norm by $|u|_{H^1(\Omega)}$ for both cases. We use similar notations for other function spaces.

For any $x_0 \in \mathbb{R}^n$ and $\delta > 0$, put

$$\mathcal{B}(x_0, \delta) \overset{\triangle}{=} \{x \in \mathbb{R}^n : |x - x_0| < \delta\}.$$

For simplicity, we use the notation $y_{x_j} = \partial_{x_j} y = \frac{\partial y}{\partial x_j}$, $j = 1, \ldots, n$ for the partial derivative of a function y with respect to x_j, where x_j is the j-th coordinate of a generic point $x = (x_1, \ldots, x_n)$ in $\mathbb{R}^n$. In a similar manner, we use the notation y_t for the partial derivative of y with respect to $t \in \mathbb{R}$.

Let us recall the classical Cauchy–Kovalevskaya theorem (e.g. [14]), which asserts the existence and uniqueness of analytic solutions to PDEs with analytic coefficients and initial data. Then, naturally, one may ask the following two questions:

1. Does there exist a non-analytic solution to the same equation?
2. Does the uniqueness of solutions still hold when the analyticity of coefficients and initial data is weakened to be differentiability?

For the above questions, when the PDE is linear, Holmgren [30] proved in 1901 that it admits at most one continuous solution with continuous partial derivatives up to the order of the equation. However, the analyticity condition is essential in Holmgren's proof. It is based on a duality argument, that is, the uniqueness follows from the existence of solutions to a suitable adjoint equation (with analytic coefficients and initial data), which is guaranteed by the Cauchy–Kovalevskaya theorem.

A fundamental contribution to remove the above analyticity condition for uniqueness was made by Carleman [11] in 1939. Let us recall this result briefly. We begin with the following notion:

A function $y \in L^2_{loc}(\mathbb{R}^n)$ is said to vanish of infinite order at $x_0 \in \mathbb{R}^n$ if there exists an $R > 0$ so that for each integer $N \in \mathbb{N}$, there is a constant $C_N > 0$ satisfying that

$$\int_{\mathcal{B}(x_0, r)} y^2 dx \leq C_N r^{2N}, \qquad \forall\, r \in (0, R).$$

Let $P = -\Delta + V$ with $V \in L^\infty_{loc}(\mathbb{R}^2)$. In [11], T. Carleman showed that *any solution $y \in H^1_{loc}(\mathbb{R}^2)$ to $Py = 0$ (in the sense of distribution) equals zero if it vanishes of infinite order at some $x_0 \in \mathbb{R}^2$*. To prove this result, he introduced a new method, now known as the Carleman estimate.

In 1954, Müller [58] extended the above method to elliptic equations on $\mathbb{R}^n$. Later, more general results on uniqueness problems for PDEs were given by A. P. Calderón in [9] and by L. Hörmander in [31, Chap. 8], by virtue of the Carleman estimate method. Since then, there are numerous publications in this area (e.g. [40, 64, 77]).

Besides applications in the study of uniqueness/unique continuation problems, people found that Carleman estimate can also be applied to solve inverse problems for PDEs (e.g. [8]). Now, Carleman estimate is one of the most useful tools in this field (e.g. [37, 41, 42, 45, 68]).

Another important application of Carleman estimate is the study of control (including controllability, observability and stabilization) problems for PDEs. Early works in this respect include [28] for the parabolic equations and [71] for the hyperbolic equations.

Most of the above mentioned control problems are strongly related to observability estimates for suitable PDEs. Various techniques, such as the spectrum method [47], the (Rellich-type) multiplier [51], the microlocal analysis [3] and the Carleman estimate (e.g. [28, 71]), have been developed to establish the desired observability estimate. Compared with other methods, the Carleman estimate-based method has its own advantage. Let us explain it briefly below.

The spectrum method works well for observability estimates for PDEs involving in some special (space) domains, i.e., intervals and rectangles, or with special coefficients (say the time-invariant case). However, it seems very hard to handle equations in general domains and/or with general coefficients.

The multiplier approach can be applied to treat observability estimates for time-reversible PDEs with time independent lower order terms. Nevertheless, it seems that it cannot be used to solve observability problems for time-irreversible PDEs or PDEs with both time and spatial dependent coefficients of lower order terms.

The microlocal analysis approach is useful in solving observability problems for many PDEs such as the wave, Schrödinger and plate equations. Further, it can give sharp sufficient conditions for the observability estimate of the wave equations. However, until now, people do not know how to apply it to equations with both time and space dependent coefficients.

Carleman estimate allows to address many observability problems (say, for PDEs with variable coefficients on general domains) which cannot be handled by the other three methods mentioned above. Further, it is robust with respect to the lower order terms and can be used to obtain explicit bounds on the observability constants/control costs in terms of the coefficients entering in the linearized equations, which are crucial in solving some controllability problems of semilinear and quasilinear PDEs. Because of this, in the recent 20 years there are many works addressed to control problems for PDEs by means of Carleman estimates (e.g. [5, 10, 16–22, 34, 43, 44, 46, 52, 53, 55, 61, 63, 65, 70, 72–76]).

In this book, we shall mainly focus on Carleman estimates for three typical PDEs, that is, elliptic, parabolic and hyperbolic equations, and especially their applications in control problems (and also unique continuation and inverse problems). As we shall see later, all of these results can be deduced from a fundamental weighted identity for second order PDOs, to be established in the last section of this chapter.

1.2 Two Stimulating Examples

In this section, we explain the key idea of Carleman estimates by two simple examples.

Stimulating example 1, Stability estimate for linear ODEs: Let $a \in L^\infty(0, T)$. Consider the following ODE:

$$\begin{cases} \dfrac{dx(t)}{dt} = a(t)x(t), \quad t \in [0, T], \\ x(0) = x_0 \in \mathbb{R}^n. \end{cases} \tag{1.1}$$

The following simple result is well-known and easy to be proved. Here we shall give a slightly different proof, which includes the main idea of Carleman estimate.

Proposition 1.1 *There is a constant $C > 0$ such that*

$$\max_{t \in [0, T]} |x(t)| \leq C|x_0|, \qquad \forall \, x_0 \in \mathbb{R}^n. \tag{1.2}$$

Proof For any $\lambda \in \mathbb{R}$, set

$$\ell(t) = -\lambda t, \quad \theta(t) = e^{\ell(t)}, \quad y(t) = \theta(t)x(t).$$

It is easy to see that

$$\theta(t)\frac{dx(t)}{dt} = \frac{dy(t)}{dt} + \lambda y(t).$$

Then

$$\theta(t)\frac{dx(t)}{dt} \cdot (2y(t)) = 2y(t)\frac{dy(t)}{dt} + 2\lambda|y(t)|^2 = \frac{d}{dt}\big(|y(t)|^2\big) + 2\lambda|y(t)|^2. \tag{1.3}$$

Combining (1.3) and the first equation of (1.1), and noting that $y(t) = \theta(t)x(t)$, we obtain that

$$\begin{aligned} \frac{d}{dt}\big(|\theta(t)x(t)|^2\big) &= \frac{d}{dt}\big(|y(t)|^2\big) \\ &= -2\lambda|y(t)|^2 + 2y(t)\theta(t)\frac{dx(t)}{dt} = 2(a(t) - \lambda)|y(t)|^2. \end{aligned} \tag{1.4}$$

Choosing λ large enough such that $a(t) - \lambda \leq 0$ (for a.e. $t \in (0, T)$) in (1.4), one finds

$$|x(t)| \leq e^{\lambda T}|x_0|, \quad \forall t \in [0, T],$$

which proves (1.2). $\qquad\qquad\qquad\qquad\qquad\qquad\qquad\qquad\qquad\qquad\qquad\quad\square$

Remark 1.1 Identity (1.3) can be re-written as

$$\frac{dx(t)}{dt} \cdot (2\theta(t)y(t)) = \frac{d}{dt}\big(|y(t)|^2\big) + 2\lambda|y(t)|^2. \tag{1.5}$$

Note that $\frac{dx(t)}{dt}$ is the principal part of the first equation in (1.1). The main idea of (1.5) is to derive a pointwise identity (and/or estimate) on the principal part $\frac{dx(t)}{dt}$ in terms of the sum of a "divergence" term $\frac{d}{dt}(|y(t)|^2)$ and an "energy" term $2\lambda|y(t)|^2$.

As we have seen in the proof of Proposition 1.1, one can choose λ to be large enough to absorb the undesired terms, which is the key of all Carleman estimates.

Stimulating example 2, Carleman estimate for the first order PDOs: For any fixed $\gamma_0 \in C(\overline{\Omega})$ and $\gamma \in [C^1(\overline{\Omega})]^n$, consider the following first order PDO:

$$\mathscr{P}(x, D) = \gamma(x) \cdot \nabla + \gamma_0(x), \quad x \in \Omega.$$

For any $x_0 \in \mathbb{R}^n \setminus \overline{\Omega}$, set

$$\phi(x) = |x - x_0|^2. \tag{1.6}$$

Proposition 1.2 *Assume that*

$$\gamma(x) \cdot (x - x_0) \leq -c_0, \quad in \ \overline{\Omega} \tag{1.7}$$

for some constant $c_0 > 0$. Then there exist two constants $\lambda^* > 0$ and $C > 0$ such that

$$\lambda \int_\Omega u^2 e^{2\lambda\phi} dx \leq C \int_\Omega |e^{\lambda\phi} \mathscr{P}(x, D)u|^2 dx, \quad \forall \lambda \geq \lambda^*, \ u \in H_0^1(\Omega). \tag{1.8}$$

Proof For $\lambda > 0$, we set

$$\ell(x) = \lambda\phi(x), \quad \theta = \theta(x) = e^{\ell(x)}, \quad v = \theta u, \tag{1.9}$$

where ϕ is given by (1.6). It is easy to see that

$$\theta \mathscr{P}(x, D)u = \theta[\gamma \cdot \nabla(\theta^{-1}v) + \gamma_0\theta^{-1}v] = \gamma \cdot \nabla v - (\gamma \cdot \nabla\ell)v + \gamma_0 v.$$

Then, a short calculation gives

$$\begin{aligned} &\theta \mathscr{P}(x, D)u \cdot (2v) \\ &= \nabla \cdot (\gamma(x)v^2) + \{-\nabla \cdot \gamma(x) - 4\lambda[\gamma(x) \cdot (x - x_0)] + 2\gamma_0\}v^2. \end{aligned} \tag{1.10}$$

By (1.6), (1.7), (1.9) and (1.10), and noting that $\gamma_0 \in C(\overline{\Omega})$ and $\gamma \in [C^1(\overline{\Omega})]^n$, we find that, for $C = |\gamma_0|_{C(\overline{\Omega})} + |\gamma|_{[C^1(\overline{\Omega})]^n}$,

$$\mathscr{P}(x, D)u \cdot (2\theta v) \geq \nabla \cdot (\gamma(x)v^2) + 2(2\lambda c_0 - C)v^2.$$

Finally, integrating the above inequality in Ω, noting that $u \in H_0^1(\Omega)$ and $v = \theta u$, choosing $\lambda^* > 0$ such that $\lambda^* c_0 > C$, then for all $\lambda \geq \lambda^*$, we have the inequality (1.8).

Remark 1.2 Similar to Proposition 1.1, the key point in the proof of Proposition 1.2 is the pointwise identity (1.10). Meanwhile, Proposition 1.2 is essentially the same as [6, Lemma 2.1], and the later is useful in the proof of logarithmic stability for inverse acoustic wave problem.

From the above two examples, one can see that a key step in deriving a Carleman estimate is to choose an appropriate weight function θ and establish a suitable identity, which contains the principal operator of the underlying equation, some divergence terms and some energy terms with a good sign. In Proposition 1.1, the weight function θ depends only on the time variable t; while in Proposition 1.2, the weight function θ depends only on the space variable x. Generally speaking, the choice of θ should depend on the PDO itself (e.g. [32, Sect. 8.6]).

1.3   A Fundamental Weighted Identity

As mentioned before, a crucial tool that we shall employ in this book is an elementary pointwise weighted identity for second order PDOs, to be presented below. This identity was stimulated by [39, 45] and established in [22, 23] (see [49] for an earlier result).

For $z(= z(t, x)) \in C^2(\mathbb{R}^{1+m}; \mathbb{C})$, we introduce a formally second order PDO $\mathscr{P}$ as:

$$\mathscr{P}z \triangleq (\alpha + i\beta)z_t + \sum_{j,k=1}^{m} \left(a^{jk}z_{x_j}\right)_{x_k}. \tag{1.11}$$

Here α, $\beta \in C^1(\mathbb{R}^{1+m}; \mathbb{R})$, and $a^{jk} \in C^1(\mathbb{R}^{1+m}; \mathbb{R})$ satisfying the following condition:

$$a^{jk} = a^{kj}, \quad j, k = 1, 2, \ldots, m. \tag{1.12}$$

Fix a weight function $\ell \in C^2(\mathbb{R}^{1+m}; \mathbb{R})$, and put

$$\theta = e^\ell, \quad v = \theta z. \tag{1.13}$$

In order to have more flexibility in the sequel, we also introduce two auxillary functions $\Psi \in C^1(\mathbb{R}^{1+m}; \mathbb{R})$ and $\Phi \in C(\mathbb{R}^{1+m}; \mathbb{R})$. Some elementary calculations yield that

$$\theta\mathscr{P}z = (\alpha + i\beta)(v_t - \ell_t v) + \sum_{j,k=1}^{m} (a^{jk}v_{x_j})_{x_k} - 2\sum_{j,k=1}^{m} a^{jk}\ell_{x_j}v_{x_k}$$
$$+ \sum_{j,k=1}^{m} a^{jk}\ell_{x_j}\ell_{x_k}v - \sum_{j,k=1}^{m} (a^{jk}\ell_{x_j})_{x_k}v$$
$$= I_1 + I_2,$$

where

$$
\begin{cases}
I_1 \overset{\triangle}{=} i\beta v_t - \alpha \ell_t v + \sum_{j,k=1}^{m} (a^{jk} v_{x_j})_{x_k} + Av, \\
I_2 \overset{\triangle}{=} \alpha v_t - i\beta \ell_t v - 2 \sum_{j,k=1}^{m} a^{jk} \ell_{x_j} v_{x_k} + \Psi v + \Phi v, \\
A \overset{\triangle}{=} \sum_{j,k=1}^{m} a^{jk} \ell_{x_j} \ell_{x_k} - \sum_{j,k=1}^{m} (a^{jk} \ell_{x_j})_{x_k} - \Psi - \Phi.
\end{cases}
\tag{1.14}
$$

We have the following fundamental identity for the operator $\mathscr{P}$.

Theorem 1.1 *It holds that*

$$
2\mathrm{Re}\,(\theta \mathscr{P} z \overline{I_1}) + M_t + \mathrm{div}\, V
$$

$$
= |I_1|^2 + |I_1 + \Phi v|^2 + B|v|^2 - 2\sum_{j,k=1}^{m} a^{jk} \Psi_{x_j} \mathrm{Re}\,(v_{x_k} \overline{v}) + 2\sum_{j,k=1}^{m} c^{jk} \mathrm{Re}\,(v_{x_j} \overline{v}_{x_k})
$$

$$
-2\sum_{j,k=1}^{m} \left[(a^{jk}(\beta \ell_t)_{x_j} + (\beta a^{jk} \ell_{x_j})_t\right] \mathrm{Im}\,(\overline{v}_{x_k} v) - 2\sum_{j,k=1}^{m} a^{jk} \alpha_{x_k} \mathrm{Re}\,(v_{x_j} \overline{v}_t) \tag{1.15}
$$

$$
-2\left[\beta \Psi + \sum_{j,k=1}^{m} (\beta a^{jk} \ell_{x_j})_{x_k}\right] \mathrm{Im}\,(\overline{v} v_t),
$$

where

$$
\begin{cases}
M \overset{\triangle}{=} \left[(\alpha^2 + \beta^2)\ell_t - \alpha A\right] |v|^2 + \alpha \sum_{j,k=1}^{m} a^{jk} v_{x_j} \overline{v}_{x_k} - 2\sum_{j,k=1}^{m} \mathrm{Im}\,(\beta a^{jk} \ell_{x_j} \overline{v}_{x_k} v), \\
V \overset{\triangle}{=} [V^1, \ldots, V^k, \ldots, V^m], \\
V^k \overset{\triangle}{=} 2\sum_{j=1}^{m} \Big\{ \sum_{j',k'=1}^{m} \left(2 a^{jk'} a^{j'k} - a^{jk} a^{j'k'}\right) \ell_{x_j} \mathrm{Re}\,(v_{x_{j'}} \overline{v}_{x_{k'}}) \\
\qquad\qquad - a^{jk}\big[\alpha \mathrm{Re}\,(v_{x_j} \overline{v}_t) - \beta\left(\ell_{x_j} \mathrm{Im}\,(\overline{v}_t v) + \ell_t \mathrm{Im}\,(v_{x_j} \overline{v})\right) \\
\qquad\qquad + \Psi \mathrm{Re}\,(v_{x_j} \overline{v}) - (A\ell_{x_j} - \alpha \ell_{x_j} \ell_t)|v|^2\big]\Big\}, \qquad k = 1, 2 \cdots, m,
\end{cases}
$$

and

$$
\begin{cases}
B \overset{\triangle}{=} (\alpha^2 \ell_t)_t + (\beta^2 \ell_t)_t - (\alpha A)_t - 2\left[\sum_{j,k=1}^{m} (\alpha a^{jk} \ell_{x_j} \ell_t)_{x_k} + \alpha \Psi \ell_t\right] \\
\qquad + 2\left[\sum_{j,k=1}^{m} (a^{jk} \ell_{x_j} A)_{x_k} + A\Psi\right] - \Phi^2 \\
c^{jk} \overset{\triangle}{=} \sum_{j',k'=1}^{m} \left[2(a^{j'k} \ell_{x_{j'}})_{x_{k'}} a^{jk'} - (a^{jk} a^{j'k'} \ell_{x_{j'}})_{x_{k'}}\right] + \frac{1}{2}(\alpha a^{jk})_t - a^{jk} \Psi.
\end{cases}
$$

Proof Recalling (1.14) for I_1 and I_2, we have

$$2\mathrm{Re}\,(\theta\mathscr{P}z\overline{I_1}) = 2|I_1|^2 + 2\mathrm{Re}\,(I_1\overline{I_2}). \qquad (1.16)$$

Let us compute $2\mathrm{Re}\,(I_1\overline{I_2})$. Denote the terms in the right hand sides of I_1 and I_2 by I_1^l ($l = 1, 2, 3, 4$) and I_2^d ($d = 1, 2, 3, 4, 5$), respectively. Then

$$2\mathrm{Re}\,(I_1\overline{I_2}) = 2\mathrm{Re}\left(I_1\overline{(I_2^1 + I_2^2 + I_2^3 + I_2^4)}\right) + 2\mathrm{Re}\left(I_1(\varPhi\overline{v})\right). \qquad (1.17)$$

Next, by (1.14), we have

$$2\mathrm{Re}\left[I_1^1(\overline{I_2^1} + \overline{I_2^2})\right] = -(\beta^2\ell_t|v|^2)_t + (\beta^2\ell_t)_t|v|^2. \qquad (1.18)$$

Noticing that $\mathrm{Re}\,(ic) = -\mathrm{Im}\,c = \mathrm{Im}\,(\overline{c})$, for any $c \in \mathbb{C}$, we have

$$\begin{aligned}
2\mathrm{Re}\left[I_1^1(\overline{I_2^3} + \overline{I_2^4})\right] &= 2\mathrm{Re}\left[i\beta v_t\left(-2\sum_{j,k=1}^{m} a^{jk}\ell_{x_j}\overline{v}_{x_k} + \varPsi\overline{v}\right)\right] \\
&= -2\mathrm{Im}\left[\beta v_t\left(-2\sum_{j,k=1}^{m} a^{jk}\ell_{x_j}\overline{v}_{x_k} + \varPsi\overline{v}\right)\right] \qquad (1.19) \\
&= 2\sum_{j,k=1}^{m}\mathrm{Im}\left(\beta a^{jk}\ell_{x_j}v\overline{v}_{x_k}\right)_t - 2\sum_{j,k=1}^{m}\mathrm{Im}\left(\beta a^{jk}\ell_{x_j}v\overline{v}_t\right)_{x_k} \\
&\quad -2\sum_{j,k=1}^{m}\left(\beta a^{jk}\ell_{x_j}\right)_t\mathrm{Im}\left(v\overline{v}_{x_k}\right) - 2\left[\beta\varPsi + \sum_{j,k=1}^{m}\left(\beta a^{jk}\ell_{x_j}\right)_{x_k}\right]\mathrm{Im}\,(\overline{v}v_t),
\end{aligned}$$

where we have used the following two facts:

$$\begin{aligned}
4\mathrm{Im}&\left(\beta v_t\sum_{j,k=1}^{m} a^{jk}\ell_{x_j}\overline{v}_{x_k}\right) \\
&= 2\mathrm{Im}\left(\beta v_t\sum_{j,k=1}^{m} a^{jk}\ell_{x_j}\overline{v}_{x_k}\right) + 2\sum_{j,k=1}^{m}\mathrm{Im}\left[\left(\beta a^{jk}\ell_{x_j}\overline{v}_{x_k}v\right)_t - \left(\beta a^{jk}\ell_{x_j}\right)_t\overline{v}_{x_k}v\right] \\
&\quad -2\sum_{j,k=1}^{m}\mathrm{Im}\left[\left(\beta a^{jk}\ell_{x_j}\overline{v}_t v\right)_{x_k} - \left(\beta a^{jk}\ell_{x_j}\right)_{x_k}\overline{v}_t v\right] + 2\mathrm{Im}\left[\beta\overline{v}_t\sum_{j,k=1}^{m} a^{jk}\ell_{x_j}v_{x_k}\right]
\end{aligned}$$

and

$$\mathrm{Im}\left(\beta\overline{v}_t\sum_{j,k=1}^{m} a^{jk}\ell_{x_j}v_{x_k}\right) = -\mathrm{Im}\left(\beta v_t\sum_{j,k=1}^{m} a^{jk}\ell_{x_j}\overline{v}_{x_k}\right).$$

Next,

$$2\mathrm{Re}\left[I_1^2\left(I_2^1 + I_2^2 + I_2^3 + I_2^4\right)\right]$$
$$= -\left(\alpha^2 \ell_t |v|^2\right)_t + 2\sum_{j,k=1}^{m}\left(\alpha a^{jk}\ell_{x_j}\ell_t|v|^2\right)_{x_k} \tag{1.20}$$
$$+\left(\alpha^2\ell_t\right)_t|v|^2 - 2\left[\sum_{j,k=1}^{m}\left(\alpha a^{jk}\ell_{x_j}\ell_t\right)_{x_k} + \alpha\Psi\ell_t\right]|v|^2.$$

By (1.12), it is easy to see that $\displaystyle\sum_{j,k=1}^{m}\left[\left(\alpha a^{jk}v_{x_j}\overline{v}_{x_k}\right)_t - \left(\alpha a^{jk}\right)_t v_{x_j}\overline{v}_{x_k}\right]$ is real-valued,

and equals to $2\mathrm{Re}\displaystyle\sum_{j,k=1}^{m}\left(\alpha a^{jk}v_{x_j}\overline{v}_{x_k t}\right)$. Hence

$$2\mathrm{Re}\left[I_1^3\left(\overline{I_2^1} + \overline{I_2^2}\right)\right]$$
$$=2\sum_{j,k=1}^{m}\mathrm{Re}\left(\alpha a^{jk}v_{x_j}\overline{v}_t\right)_{x_k} - 2\sum_{j,k=1}^{m}\mathrm{Re}\left(a^{jk}\alpha_{x_k}v_{x_j}\overline{v}_t\right) - \sum_{j,k=1}^{m}\left[\left(\alpha a^{jk}v_{x_j}\overline{v}_{x_k}\right)_t - \left(\alpha a^{jk}\right)_t v_{x_j}\overline{v}_{x_k}\right]$$
$$-2\sum_{j,k=1}^{m}\mathrm{Im}\left[\left(\beta a^{jk}\ell_t v_{x_j}\overline{v}\right)_{x_k} + a^{jk}\left(\beta\ell_t\right)_{x_k}\overline{v}_{x_j}v\right] \tag{1.21}$$
$$= 2\sum_{j,k=1}^{m}\left[\alpha a^{jk}\mathrm{Re}\left(v_{x_j}\overline{v}_t\right) - \beta a^{jk}\ell_t\mathrm{Im}\left(v_{x_j}\overline{v}\right)\right]_{x_k} - \sum_{j,k=1}^{m}\left(\alpha a^{jk}v_{x_j}\overline{v}_{x_k}\right)_t$$
$$-2\sum_{j,k=1}^{m}a^{jk}\alpha_{x_k}\mathrm{Re}\left(v_{x_j}\overline{v}_t\right) + \sum_{j,k=1}^{m}\left(\alpha a^{jk}\right)_t v_{x_j}\overline{v}_{x_k} - 2\sum_{j,k=1}^{m}a^{jk}\left(\beta\ell_t\right)_{x_k}\mathrm{Im}\left(\overline{v}_{x_j}v\right).$$

Using the condition (1.12) again, we obtain that

$$4\sum_{j,k,j',k'=1}^{m}a^{jk}a^{j'k'}\ell_{x_j}\mathrm{Re}\left(v_{x_{j'}}\overline{v}_{x_k x_{k'}}\right)$$
$$= 2\sum_{j,k,j',k'=1}^{m}a^{jk}a^{j'k'}\ell_{x_j}\left(v_{x_{j'}}\overline{v}_{x_k x_{k'}} + \overline{v}_{x_{j'}}v_{x_k x_{k'}}\right) \tag{1.22}$$
$$= \sum_{j,k,j',k'=1}^{m}\left\{\left[a^{jk}a^{j'k'}\ell_{x_j}\left(v_{x_{j'}}\overline{v}_{x_{k'}} + \overline{v}_{x_{j'}}v_{x_{k'}}\right)\right]_{x_k} - \left(a^{jk}a^{j'k'}\ell_{x_j}\right)_{x_k}\left(v_{x_{j'}}\overline{v}_{x_{k'}} + \overline{v}_{x_{j'}}v_{x_{k'}}\right)\right\}$$
$$= 2\sum_{j,k,j',k'=1}^{m}\left[a^{jk}a^{j'k'}\ell_{x_j}\mathrm{Re}\left(v_{x_{j'}}\overline{v}_{x_{k'}}\right)\right]_{x_k} - 2\sum_{j,k,j',k'=1}^{m}\left(a^{jk}a^{j'k'}\ell_{x_j}\right)_{x_k}\mathrm{Re}\left(v_{x_{j'}}\overline{v}_{x_{k'}}\right).$$

It follows from (1.22) that

$$2\mathrm{Re}\,(I_1^3 \overline{I_2^3})$$
$$= -4\mathrm{Re}\sum_{j,k,j',k'=1}^{m}\left[\left(a^{jk}\ell_{x_j}a^{j'k'}v_{x_{j'}}\overline{v}_{x_k}\right)_{x_{k'}} - a^{j'k'}(a^{jk}\ell_{x_j})_{x_{k'}}v_{x_{j'}}\overline{v}_{x_k}\right]$$
$$+2\mathrm{Re}\sum_{j,k,j',k'=1}^{m}\left[\left(a^{jk}a^{j'k'}\ell_{x_j}v_{x_{j'}}\overline{v}_{x_{k'}}\right)_{x_k} - (a^{jk}a^{j'k'}\ell_{x_j})_{x_k}v_{x_{j'}}\overline{v}_{x_{k'}}\right].$$
(1.23)

Further,

$$2\mathrm{Re}\,(I_1^3 \overline{I_2^4})$$
$$= 2\sum_{j,k=1}^{m}\mathrm{Re}\left(\Psi a^{jk}v_{x_j}\overline{v}\right)_{x_k} - 2\Psi\sum_{j,k=1}^{m}a^{jk}v_{x_j}\overline{v}_{x_k} - 2\sum_{j,k=1}^{m}a^{jk}\Psi_{x_k}\mathrm{Re}\,(v_{x_j}\overline{v}).$$
(1.24)

Finally,

$$2\mathrm{Re}\,(I_1^4(\overline{I_2^1 + I_2^2 + I_2^3 + I_2^4}))$$
$$= (\alpha A|v|^2)_t - (\alpha A)_t|v|^2 - 2\sum_{j,k=1}^{m}(a^{jk}\ell_{x_j}A|v|^2)_{x_k}$$
$$+2\left[\sum_{j,k=1}^{m}(a^{jk}\ell_{x_j}A)_{x_k} + A\Psi\right]|v|^2.$$
(1.25)

Now, by (1.17)–(1.25), combining all "$\frac{\partial}{\partial t}$"-terms, all "$\frac{\partial}{\partial x_k}$"-terms, and (1.16), we arrive at the desired identity (1.15). This completes the proof of Theorem 1.1.

Several remarks are in order.

Remark 1.3 In this subsection, for simplicity, we assume that $a^{jk} \in C^1(\mathbb{R}^{1+m}; \mathbb{R})$ ($j, k = 1, \ldots, 1 + m$). From the proof of Theorem 1.1, one can see that this condition can be relaxed to that a^{jk} is Lipschitz continuous for any $j, k = 1, \ldots, 1 + m$. In the later case, (1.15) holds for a.e. $x \in \mathbb{R}^{1+m}$. But this is enough for our purpose since we shall integrate (1.15) to derive the desired Carleman estimate. More details can be found in the rest of this book.

Remark 1.4 In Theorem 1.1, if $\alpha \equiv 1$, $\beta \equiv 0$ and z is real-valued, then the identity (1.15) is specialized as as

$$2\theta\mathscr{P}zI_1 + M_t + \mathrm{div}\,V$$
$$= |I_1|^2 + |I_1 + \Phi v|^2 + B|v|^2 - 2\sum_{j,k=1}^{m}a^{jk}\Psi_{x_j}v_{x_k}v + 2\sum_{j,k=1}^{m}c^{jk}v_{x_j}v_{x_k}.$$
(1.26)

It is easy to see that the last two terms in the left hand side of (1.26) are the "divergence"-terms, while the last three terms in the right hand side of (1.26) are the "energy"-terms and the lower order terms.

Remark 1.5 Theorem 1.1 is very much like [23, Theorem 2.1]. The main difference is the regularity of Ψ. In [23], Ψ was required to be in $C^2(\mathbb{R}^{1+m}; \mathbb{R})$. Here we relax the restriction of Ψ to be in $C^1(\mathbb{R}^{1+m}; \mathbb{R})$, by means of a small modification of the proof, i.e., keeping the term "$-2\sum\limits_{j,k=1}^{m} a^{jk}\Psi_{x_j} \mathrm{Re}\,(v_{x_k}\overline{v})$" in the right hand side of (1.15), which can be finally absorbed by the "energy"-terms. Usually, when applying Theorem 1.1 to concrete problems, the choice of Ψ depends on the coefficients of the principal part of the underlying equations. In this way, we may relax the regularity conditions on the coefficients in the problems under consideration.

Remark 1.6 We see that only the symmetry condition of $(a^{jk})_{m\times m}$ is assumed in Theorem 1.1. Therefore, this theorem is applicable to ultra-hyperbolic, ultra-parabolic differential operators and degenerate elliptic/parabolic/hyperbolic equations.

Remark 1.7 If $a^{jk}(t, x) \equiv a^{jk}(x)$ and $\alpha(t, x) = \beta(t, x) \equiv 0$ in Theorem 1.1, one obtains a pointwise identity for second order elliptic operators. Based on this, we can derive some Carleman estimates for the elliptic operators. As applications, we may obtain the unique continuation, observability estimate and some other related results presented in [2, 7, 13, 15, 24–26, 29, 38, 47, 48, 54, 56, 60]. Some results in this respect will be presented in Chap. 2.

Remark 1.8 By choosing $\alpha(t, x) \equiv 1$, $\beta(t, x) \equiv 0$ and a suitable auxiliary function Ψ (see (3.12)) in Theorem 1.1, one obtains a weighted identity for second order parabolic operators. Based on this, one may recover the known results on the unique continuation, controllability/observability and inverse problems for parabolic equations in [12, 16, 19, 20, 28, 33, 36, 50, 52, 62, 66–68]. Some of the details will be provided in Chap. 3.

Remark 1.9 By choosing $\alpha(t, x) = \beta(t, x) \equiv 0$, $m = 1 + n$ and

$$(a^{jk}(t, x))_{1\leq j,k\leq m} \equiv \begin{pmatrix} 1 & 0 \\ 0 & -(b^{jk}(x))_{1\leq j,k\leq n} \end{pmatrix}$$

with $(b^{jk}(x))_{1\leq j,k\leq n} \in C^2(\mathbb{R}^n; \mathbb{R}^{n\times n})$ in Theorem 1.1, one obtains a pointwise identity for second order hyperbolic operators. Based on this, one can recover the known results on the unique continuation, controllability/observability and inverse problems for the general hyperbolic equations in [5, 8, 17, 27, 35, 44, 72]. Some of the details for these results will be presented in Chap. 4.

Remark 1.10 In the rest of this book, we shall always choose $\beta \equiv 0$ in Theorem 1.1. Nevertheless, in this chapter we provide the general complex form of this result because the readers may apply it to some other types of PDEs, say the Schrödinger and the plate equations. For example, by choosing $(a^{jk})_{1\leq j,k\leq m}$ to be the identity matrix, $\alpha(t, x) \equiv 0$, $\beta(t, x) \equiv 1$ and $\Psi = -\Delta\ell$ in Theorem 1.1, one obtains the pointwise identity derived in [43] for observability results for the nonconservative Schrödinger equations. Also, this yields the controllability/observability results in [74] for the plate equation and the results for the inverse problem of Schrödinger equation in [4, 57]. Further, letting $(a^{jk})_{1 \le j,k \le m}$ be the identity matrix, $\alpha(t, x) \equiv 0$, $\beta(t, x) \equiv p(x)$, $\Psi = -\Delta \ell$ and $\Phi \equiv 0$, one obtains the pointwise identity for the Schrödinger operator $ip(x)\partial_t + \Delta$. Then, by choosing

$$\ell(t, x) = s\varphi, \quad \varphi = e^{\gamma(|x-x_0|^2 - c|t-t_0|^2)}$$

with some constants $\gamma > 0$ and $c > 0$, and $x_0 \in \mathbb{R}^n \setminus \overline{\Omega}$ so that

$$(\nabla \log p) \cdot (x - x_0) > -2 \quad \text{in } \Omega,$$

one may recover the Carleman estimate for the Schrödinger operator $ip(x)\partial_t + \Delta$ derived in [69, Lemma 2.1].

Remark 1.11 In the identity (1.15), we introduce the weight function $\theta = e^\ell$, which shall play key roles to establish various Carleman estimates in the rest of this book. The choice of the function ℓ depends on the type of the underlying equations. In principle, the operator $\mathscr{P}$ and the weight ℓ should satisfy the so-called strong pseudo-convexity condition introduced by L. Hörmander (e.g. [31, Definition 8.6.1] and [32, Definition 28.3.1]).

References

1. Adams, R.A.: Sobolev Spaces. Academic, Cambridge (1975)
2. Aronszajn, N.: A unique continuation theorem for solutions of elliptic partial differential equations or inequalities of second order. J. Math. Pures Appl. **36**, 235–249 (1957)
3. Bardos, C., Lebeau, G., Rauch, J.: Sharp sufficient conditions for the observation, control and stabilization of waves from the boundary. SIAM J. Control Optim. **30**, 1024–1065 (1992)
4. Baudouin, L., Puel, J.P.: Uniqueness and stability in an inverse problem for the Schrödinger equation. Inverse Probl. **18**, 1537–1554 (2002)
5. Baudouin, L., de Buhan, M., Ervedoza, S.: Global Carleman estimates for waves and applications. Commun. Part. Differ. Equ. **38**, 823–859 (2013)
6. Bellassoued, M., Yamamoto, M.: Logarithmic stability in determination of a coefficient in an acoustic equation by arbitrary boundary observation. J. Math. Pures Appl. **85**, 193–224 (2006)
7. Bourgain, J., Kenig, C.: On localization in the continuous Anderson-Bernoulli model in higher dimension. Invent. Math. **161**, 389–426 (2005)
8. Bukhgeim, A.L., Klibanov, M.V.: Global uniqueness of class of multidimensional inverse problems. Soviet Math. Dokl. **24**, 244–247 (1981)
9. Calderón, A.P.: Uniqueness in the cauchy problem for partial differential equations. Am. J. Math. **80**, 16–36 (1958)
10. Cannarsa, P., Martinez, P., Vancostenoble, J.: Global Carleman Estimates for Degenerate Parabolic Operators with Applications. Memoirs of the American Mathematical Society, vol. 239 (2016)
11. Carleman, T.: Sur un problème d'unicité pour les systèmes d'équations aux dérivées partielles à deux variables indépendantes. Ark. Mat. Astr. Fys. 26 B. **17**, 1–9 (1939)
12. Chen, X.: A strong unique continuation theorem for parabolic equations. Math. Ann. **311**, 603–630 (1998)

13. Choulli, M.: Applications of Elliptic Carleman Inequalities to Cauchy and Inverse Problems. Springer Briefs in Mathematics. BCAM Springer Briefs. Springer, Cham (2016)
14. Courant, R., Hilbert, D.: Methods of Mathematical Physics II. Wiley-Interscience, New York (1962)
15. Donnelly, H., Fefferman, C.: Nodal sets of eigenfunctions on Riemannian manifolds. Invent. Math. **93**, 161–183 (1988)
16. Doubova, A., Fernández-Cara, E., González-Burgos, M., Zuazua, E.: On the controllability of parabolic systems with a nonlinear term involving the state and the gradient. SIAM J. Control Optim. **41**, 798–819 (2002)
17. Duyckaerts, T., Zhang, X., Zuazua, E.: On the optimality of the observability inequalities for parabolic and hyperbolic systems with potentials. Ann. Inst. H. Poincaré Anal. Non Linéaire. **25**, 1–41 (2008)
18. Fernández-Cara, E., Zuazua, E.: Null and approximate controllability for weakly blowing up semilinear heat equations. Ann. Inst. H. Poincaré Anal. Non Linéaire. **17**, 583–616 (2000)
19. Fernández-Cara, E.: Null controllability of the semilinear heat equations. ESAIM: Control Optim. Calc. Var. **2**, 87–107 (1997)
20. Fernández-Cara, E., Zuazua, E.: The cost of approximate controllability for heat equations: the linear case. Adv. Differ. Equ. **5**, 465–514 (2000)
21. Fernández-Cara, E., Guerrero, S., Imanuvilov, OYu., Puel, J.-P.: Local exact controllability of the Navier-Stokes system. J. Math. Pures Appl. **83**, 1501–1542 (2004)
22. Fu, X.: A weighted identity for partial differential operators of second order and its applications. C. R. Acad. Sci. Paris Série I(342), 579–584 (2006)
23. Fu, X.: Null controllability for the parabolic equation with a complex principal part. J. Funct. Anal. **257**, 1333–1354 (2009)
24. Fu, X.: Logarithmic decay of hyperbolic equations with arbitrary small boundary damping. Commun. Partial Differ. Equ. **34**, 957–975 (2009)
25. Fu, X.: Longtime behavior of the hyperbolic equations with an arbitrary internal damping. Z. angew. Math. Phys. **62**, 667–680 (2011)
26. Fu, X.: Sharp decay rates for the weakly coupled hyperbolic system with one internal damping. SIAM J. Control Optim. **50**, 1643–1660 (2012)
27. Fu, X., Yong, J., Zhang, X.: Exact controllability for the multidimensional semilinear hyperbolic equations. SIAM J. Control Optim. **46**, 1578–1614 (2007)
28. Fursikov, A.V., Imanuvilov, O.Y.: Controllability of Evolution Equations. Lecture Notes Series, vol. 34. Seoul National University, Seoul, Korea (1996)
29. Garofalo, N., Lin, F.: Monotonicity properties of variational integrals, A_p weights and unique continuation. Indiana Univ. Math. J. **35**, 245–268 (1986)
30. Holmgren, E.: Über systeme von linearen partiellen differentialgleichungen. Öfversigt af Kongl. Vetenskaps-Academien Förhandlinger **58**, 91–103 (1901)
31. Hörmander, L.: Linear Partial Differential Operators. Springer, Berlin (1963)
32. Hörmander, L.: The Analysis of Linear Partial Differential Operators IV. Fourier Integral Operators. Springer, Berlin (1985)
33. Imanuvilov, OYu.: Controllability of the parabolic equations. Sbornik Math. **186**, 879–900 (1995)
34. Imanuvilov, OYu.: On Carlerman estimates for hyperbolic equations. Asymptot. Anal. **32**, 185–220 (2002)
35. Imanuvilov, OYu., Yamamoto, M.: Global Lipschitz stability in an inverse hyperbolic problem by interior observations. Inverse Probl. **17**, 717–728 (2001)
36. Imanuvilov, OYu., Yamamoto, M.: Carleman inequalities for parabolic equations in a Sobolev spaces of negative order and exact controllability for semilinear parabolic equations. Publ. RIMS Kyoto Univ. **39**, 227–274 (2003)
37. Isakov, V.: Inverse Problems for Partial Differential Equations. Springer, New York (2006)
38. Jerison, D., Lebeau, G.: Nodal sets of sums of eigenfunctions. In: Harmonic Analysis and Partial Differential Equations, pp. 223–239. Chicago (1996). Chicago Lectures in Mathematics, University of Chicago Press, Chicago (1999)

39. Kazemi, M., Klibanov, M.V.: Stability estimates for ill-posed Cauchy problems involving hyperbolic equations and inequalities. Appl. Anal. **50**, 93–102 (1993)
40. Kenig, C.E.: Carleman estimates, uniform Sobolev inequalities for second-order differential operators, and unique continuation theorems. In: Proceedings of the International Congress of Mathematicians, vol. II, pp. 948–960. Berkeley, California, USA (1986)
41. Klibanov, M.V.: Carleman estimates for global uniqueness, stability and numerical methods for coefficient inverse problems. J. Inverse Ill-Posed Probl. **21**, 477–560 (2013)
42. Klibanov, M.V., Timonov, A.: Carleman Estimates for Coefficient Inverse Problems and Numerical Applications. Inverse and Ill-Posed Problems Series. VSP, Utrecht (2004)
43. Lasiecka, I., Triggiani, R., Zhang, X.: Global uniqueness, observability and stabilization of nonconservative Schrödinger equations via pointwise Carleman estimates. I. $H^1(\Omega)$-estimates. J. Inverse Ill-Posed Probl. **11**, 43–123 (2004)
44. Lasiecka, I., Triggiani, R., Zhang, X.: Nonconservative wave equations with purely Neumann B. C.: global uniqueness and observability in one shot. Contemp. Math. **268**, 227–326 (2000)
45. Lavrent'ev, M.M., Romanov, V.G., Shishatskiĭ, S.P.: Ill-Posed Problems of Mathematical Physics and Analysis. Translations of Mathematical Monographs, vol. 64. American Mathematical Society, Providence (1986)
46. Le Rousseau, J., Robbiano, L.: Local and global Carleman estimates for parabolic operators with coefficients with jumps at interfaces. Invent. Math. **183**, 245–336 (2011)
47. Lebeau, G., Robbiano, L.: Contrôle exact de l'équation de la chaleur. Commun. Part. Differ. Equ. **20**, 335–356 (1995)
48. Lebeau, G., Robbiano, L.: Stabilisation de l'équation des ondes par le bord. Duke Math. J. **86**, 465–491 (1997)
49. Li, W., Zhang, X.: Controllability of parabolic and hyperbolic equations: toward a unified theory. In: Control Theory of Partial Differential Equations. Lecture Notes in Pure and Applied Mathematics, vol. 242, pp. 157–174. Chapan & Hall/CRC, Boca Raton (2005)
50. Lin, F.: A uniqueness theorem for parabolic equations. Commun. Pure Appl. Math. **43**, 127–136 (1990)
51. Lions, J.L.: Contrôlabilité Exacte, Perturbations et Stabilisation de Systèmes distribués, Tome 1, Contrôlabilité exacte. Recherches en Mathématiques Appliquées. vol. 8. Masson, Paris (1988)
52. Liu, X., Zhang, X.: Local controllability of multidimensional quasilinear parabolic equations. SIAM J. Control Optim. **50**, 2046–2064 (2012)
53. López, A., Zhang, X., Zuazua, E.: Null controllality of the heat equation as singular limit of the exact controllability of dissipative wave equations. J. Math. Pures Appl. **79**, 741–808 (2000)
54. Lü, Q.: A lower bound on local energy of partial sum of eigenfunctions for Laplace-Beltrami operators. ESAIM: Control Optim. Calc. Var. **19**, 255–273 (2013)
55. Lü, Q.: Observability estimate and state observation problems for stochastic hyperbolic equations. Inverse Probl. **29**, 095011 (2013)
56. Lü, Q., Yin, Z.: The L^∞-null controllability of parabolic equation with equivalued surface boundary conditions. Asymptot. Anal. **83**, 355–378 (2013)
57. Mercado, A., Osses, A., Rosier, L.: Inverse problems for the Schrödinger equation via Carleman inequalities with degenerate weights. Inverse Probl. **24** (2008)
58. Müller, C.: On the behavior of the solutions of the differential equation $\Delta U = F(x, U)$ in the neighborhood of a point. Commun. Pure Appl. Math. **7**, 505–515 (1954)
59. Pazy, A.: Semigroups of linear operators and applications to partial differential equations. In: Applied Mathematical Sciences, vol. 44. Springer, New York (1983)
60. Phung, K.D., Zhang, X.: Time reversal focusing of the initial state for Kirchoff plate. SIAM J. Appl. Math. **68**, 1535–1556 (2008)
61. Robbiano, L.: Carleman estimates, results on control and stabilization for partial differential equations. In: Proceedings of the International Congress of Mathematicians, vol. IV, pp. 897–919. Seoul, South Korea (2014)
62. Saut, J.C., Scheurer, B.: Unique continuation for some evolution equations. J. Differ. Equ. **66**, 118–139 (1987)

63. Tang, S., Zhang, X.: Null controllability for forward and backward stochastic parabolic equations. SIAM J. Control Optim. **48**, 2191–2216 (2009)
64. Tataru, D.: Carleman estimates and unique continuation for solutions to boundary value problems. J. Math. Pures Appl. **75**, 367–408 (1996)
65. Tebou, L.: A Carleman estimate based approach for the stabilization of some locally damped semilinear hyperbolic equations. ESAIM: Control Optim. Calc. Var. **14**, 561–574 (2008)
66. Vessella, S.: Quantitative estimates of unique continuation for parabolic equations, determination of unknown time-caring boundaries and optimal stability estimates. Inverse Probl. **24**, 023001 (2008)
67. Yamabe, H.: A unique continuation theorem of a diffusion equation. Ann. Math. **69**, 462–466 (1959)
68. Yamamoto, M.: Carleman estimates for parabolic equations and applications. Inverse Probl. **25**, 123013 (2009)
69. Yuan, G., Yamamoto, M.: Lipschitz stability in inverse problems for a Kirchhoff plate equation. Asympt. Anal. **53**, 29–60 (2007)
70. Zhang, X.: A unified controllability/observability theory for some stochastic and deterministic partial differential equations. In: Proceedings of the International Congress of Mathematicians, vol. IV, pp. 3008–3034. Hyderabad, India (2010)
71. Zhang, X.: Exact controllability of the semilinear distributed parameter system and some related problems. PhD thesis, Fudan University, Shanghai, China (1998)
72. Zhang, X.: Explicit observability estimate for the wave equation with potential and its application. R. Soc. Lond. Proc. Ser. A Math. Phys. Eng. Sci. **456**, 1101–1115 (2000)
73. Zhang, X.: Explicit observability inequalities for the wave equation with lower order terms by means of Carleman inequalities. SIAM J. Control Optim. **39**, 812–834 (2001)
74. Zhang, X.: Exact controllability of semilinear plate equations. Asymptot. Anal. **27**, 95–125 (2001)
75. Zhang, X.: Carleman and observability estimates for stochastic wave equations. SIAM J. Math. Anal. **40**, 851–868 (2008)
76. Zuazua, E.: Controllability and observability of partial differential equations: some results and open problems. In: Handbook of Differential Equations: Evolutionary Differential Equations, vol. 3, pp. 527–621. Elsevier Science, Amsterdam (2006)
77. Zuily, C.: Uniqueness and Non-Uniqueness in the Cauchy Problem. Birkhäuser, Boston (1983)

Chapter 2
Carleman Estimates for Second Order Elliptic Operators and Applications, a Unified Approach

Abstract In this chapter, we establish two Carleman estimates (with different weight functions) for second order elliptic operators, i.e. Theorems 2.1 and 2.2. By means of the first one, we derive an interpolation inequality for elliptic equations, via which an observability estimate for sums of eigenfunctions of elliptic operators and a stabilization result for locally damped hyperbolic equations are proved. Based on the second one, we show a strong unique continuation property and a three-ball inequality for elliptic equations.

Keywords Carleman estimate · Second order elliptic operator · Observability estimate · Strong unique continuation · Three-ball inequality

Throughout this chapter, we assume the following condition.

Condition 2.1 *The functions $h^{jk}(\cdot) \in C^1(\overline{\Omega}; \mathbb{R})$ (for $j, k = 1, \dots, n$) satisfy*

$$h^{jk}(x) = h^{kj}(x), \qquad \forall\, x \in \overline{\Omega}, \tag{2.1}$$

and for some constant $h_0 > 0$,

$$\sum_{j,k=1}^{n} h^{jk}(x)\xi^j \overline{\xi}^k \geq h_0|\xi|^2, \qquad \forall\, (x, \xi^1, \dots, \xi^n) \in \overline{\Omega} \times \mathbb{C}^n. \tag{2.2}$$

Remark 2.1 The above $C^1(\overline{\Omega}; \mathbb{R})$ regularity assumption for h^{jk} can be relaxed to the Lipschitz continuity. Indeed, any Lipschitz continuous function is differentiable almost everywhere with a bounded derivative function, and therefore, all the proofs in this chapter still work with some standard and small modifications.

In the rest of this chapter, unless otherwise stated, we shall denote by $C = C(\Omega, n, (h^{jk})_{1 \leq j,k \leq n})$ a generic positive constant, which may change from line to line.

© The Author(s), under exclusive license to Springer Nature Switzerland AG 2019
X. Fu et al., *Carleman Estimates for Second Order Partial Differential Operators and Applications*, SpringerBriefs in Mathematics,
https://doi.org/10.1007/978-3-030-29530-1_2

2.1   Carleman Estimates for Second Order Elliptic Operators

The purpose of this section is to establish two Carleman estimates for second order elliptic operators.

We begin with the following result, which is an immediate consequence of Theorem 1.1.

Lemma 2.1 *Let* $a^{jk} \in C^1(\mathbb{R}^m; \mathbb{R})$ *satisfy (1.12). Assume that* $z(= z(x)) \in C^2(\mathbb{R}^m; \mathbb{C})$, $\Psi \in C^1(\mathbb{R}^m; \mathbb{R})$, $\Phi \in C(\mathbb{R}^m; \mathbb{R})$ *and* $\ell \in C^2(\mathbb{R}^m; \mathbb{R})$. *Then (see (1.13) for* θ *and* v)

$$\theta^2 \left| \sum_{j,k=1}^{m} (a^{jk} z_{x_j})_{x_k} \right|^2 + \mathrm{div}\, V$$

$$\geq |I_1 + \Phi v|^2 + 2 \sum_{j,k=1}^{m} c^{jk} \mathrm{Re}\, (v_{x_j} \bar{v}_{x_k}) - 2 \sum_{j,k=1}^{m} a^{jk} \Psi_{x_j} \mathrm{Re}\, (v_{x_k} \bar{v}) + B|v|^2. \tag{2.3}$$

Here

$$\begin{cases} I_1 = \displaystyle\sum_{j,k=1}^{m} (a^{jk} v_{x_j})_{x_k} + Av, \\ A = \displaystyle\sum_{j,k=1}^{m} a^{jk} \ell_{x_j} \ell_{x_k} - \sum_{j,k=1}^{m} (a^{jk} \ell_{x_j})_{x_k} - \Psi - \Phi \end{cases} \tag{2.4}$$

and

$$\begin{cases} V = [V^1, \ldots, V^k, \ldots, V^m], \\ V^k = 2\,\mathrm{Re} \displaystyle\sum_{j=1}^{m} \Big[\sum_{j',k'=1}^{m} (2a^{jk'} a^{j'k} - a^{jk} a^{j'k'}) \ell_{x_j} v_{x_{j'}} \bar{v}_{x_{k'}} - a^{jk} \left(\Psi v_{x_j} \bar{v} - A\ell_{x_j} |v|^2 \right) \Big], \\ c^{jk} = \displaystyle\sum_{j',k'=1}^{m} [2(a^{j'k} \ell_{x_{j'}})_{x_{k'}} a^{jk'} - (a^{jk} a^{j'k'} \ell_{x_{j'}})_{x_{k'}}] - a^{jk} \Psi, \\ B = 2 \displaystyle\sum_{j,k=1}^{m} \left(a^{jk} \ell_{x_j} A \right)_{x_k} + 2A\Psi - \Phi^2. \end{cases} \tag{2.5}$$

Proof We choose $\alpha = \beta = 0$ in Theorem 1.1. Noting that

$$2\,\mathrm{Re}\,(\theta \mathscr{P} z \bar{I_1}) \leq \theta^2 |\mathscr{P} z|^2 + |I_1|^2,$$

by some elementary calculations, we obtain the desired inequality (2.3) immediately. This completes the proof of Lemma 2.1.

Remark 2.2 In the rest of this chapter, we shall establish two Carleman estimates for the second order elliptic partial differential operator by choosing two different weight functions, respectively. Generally speaking, when the desired weight function (for the corresponding Carleman estimate) exists, it may not be unique. Nevertheless, any candidate of weight functions should satisfy the so-called strongly pseudo-convex condition (e.g. [17, Definition 8.6.1] and [19, Definition 28.3.1]).

Fix $b_1, b_2 \in \mathbb{R}$ with $b_1 < b_2$. Choose a function $\varphi(= \varphi(s, x)) \in C^2([b_1, b_2] \times \overline{\Omega}))$ (which will be given later) satisfying

$$\varphi_{sx_j} = 0, \quad j = 1, 2, \ldots, n, \tag{2.6}$$

and put

$$\ell(s, x) = \lambda\phi(s, x), \quad \phi(s, x) = e^{\mu\varphi(s,x)}, \tag{2.7}$$

where $\lambda, \mu > 1$ are two parameters. For $k \in \mathbb{N}$, we denote by $O(\lambda^k)$ a function of order λ^k for large λ.

Our first Carleman estimate is as follows.

Theorem 2.1 *Let φ satisfy (2.6), and ℓ and ϕ be given in (2.7). Then, there is a constant $\mu_1 > 0$ such that for all $\mu \geq \mu_1$, one can find two constants $C = C(\mu) > 0$ and $\lambda_1 = \lambda_1(\mu)$ so that for all $z(= z(s, x)) \in C^2([b_1, b_2] \times \overline{\Omega}; \mathbb{C})$ with $z(b_1, \cdot) = z(b_2, \cdot) = z_s(b_1, \cdot) = z_s(b_2, \cdot) = 0$ and $\lambda \geq \lambda_1$, it holds that*

$$\lambda\mu^2 \int_{b_1}^{b_2} \int_\Omega \theta^2 \phi |\nabla\varphi|^2 \left(|z_s|^2 + |\nabla z|^2 + \lambda^2\mu^2\phi^2|\nabla\varphi|^2|z|^2\right) dxds$$

$$\leq C\Bigg[\int_{b_1}^{b_2} \int_\Omega \theta^2 \left| z_{ss} + \sum_{j,k=1}^n (h^{jk} z_{x_j})_{x_k} \right|^2 dxds + \int_{b_1}^{b_2} \int_\Gamma \sum_{k=1}^n V^k v^k dxds \tag{2.8}$$

$$+ \lambda\mu \int_{b_1}^{b_2} \int_\Omega \theta^2 \phi \left(|z_s|^2 + |\nabla z|^2 + \lambda^2\mu^2\phi^2|z|^2\right) dxds \Bigg],$$

where θ and v are given in (1.13),

$$\begin{cases} V^k = 2\,\mathrm{Re}\, \sum_{j=1}^n \left[h^{jk}\left(2\ell_s v_{x_j}\bar{v}_s - \ell_{x_j}|v_s|^2 - \Psi v_{x_j}\bar{v} + A\ell_{x_j}|v|^2\right) \right. \\ \qquad\qquad \left. + \sum_{j',k'=1}^n \left(2h^{jk'}h^{j'k} - h^{jk}h^{j'k'}\right)\ell_{x_j} v_{x_{j'}}\bar{v}_{x_{k'}} \right], \quad k = 1, 2, \ldots, n, \\ A = \sum_{j,k=1}^n h^{jk}\ell_{x_j}\ell_{x_k} - \sum_{j,k=1}^n (h^{jk}\ell_{x_j})_{x_k} + \ell_s^2 - \ell_{ss} - \Psi - \Phi, \end{cases}$$

$$\tag{2.9}$$

and

$$\Psi = -2\lambda\mu^2\phi\Big(|\varphi_s|^2 + \sum_{j,k=1}^{n} h^{jk}\varphi_{x_j}\varphi_{x_k}\Big), \quad \Phi = -\Psi - \sum_{j,k=1}^{n} (h^{jk}\ell_{x_j})_{x_k} - \ell_{ss}.$$
(2.10)

Proof We divide the proof into three steps.

Step 1. A pointwise inequality. In Lemma 2.1, we choose $m = 1 + n$ and

$$(a^{jk})_{m\times m} = \begin{pmatrix} 1 & 0 \\ 0 & (h^{jk})_{n\times n} \end{pmatrix}.$$

Then, it follows from (2.3) and (2.5) that

$$\theta^2\Big|z_{ss} + \sum_{j,k=1}^{n} (h^{jk}z_{x_j})_{x_k}\Big|^2 + M_s + \text{div } V$$

$$\geq 2\Big(\ell_{ss} - \sum_{j,k=1}^{n} (h^{jk}\ell_{x_j})_{x_k} - \Psi\Big)|v_s|^2 + 2\sum_{j,k=1}^{n} c^{jk}\,\text{Re}\,(v_{x_j}\bar{v}_{x_k}) \qquad (2.11)$$

$$+8\sum_{j,k=1}^{n} h^{jk}\ell_{x_js}\,\text{Re}\,(v_{x_k}\bar{v}_s) - 2\sum_{j,k=1}^{n} h^{jk}\Psi_{x_j}\,\text{Re}\,(v_{x_k}\bar{v}) - 2\Psi_s\,\text{Re}\,(v_s\bar{v}) + B|v|^2.$$

Here

$$\begin{cases} M = 2\ell_s\Big(|v_s|^2 - \sum_{j,k=1}^{n} h^{jk}v_{x_j}\bar{v}_{x_k} + A|v|^2\Big) \\ \qquad\qquad + 4\sum_{j,k=1}^{n} h^{jk}\ell_{x_j}\,\text{Re}\,(v_{x_k}\bar{v}_s) - 2\Psi\,\text{Re}\,(v_s\bar{v}), \\ c^{jk} = \sum_{j',k'=1}^{n}\big[2(h^{j'k}\ell_{x_{j'}})_{x_{k'}}h^{jk'} - (h^{jk}h^{j'k'}\ell_{x_{j'}})_{x_{k'}}\big] - h^{jk}\ell_{ss} - h^{jk}\Psi, \\ \qquad\qquad\qquad\qquad\qquad\qquad\qquad\qquad\qquad j,k = 1,2,\ldots,n, \\ B = 2(A\ell_s)_s + 2\sum_{j,k=1}^{n} (h^{jk}\ell_{x_j}A)_{x_k} + 2A\Psi - \Phi^2, \end{cases}$$
(2.12)

where Ψ and Φ are given in (2.10), and A is given in (2.9).

Step 2. Estimation of the "energy-terms" (in the right-hand side of (2.11)). By (2.7) and noting (2.6), it is easy to see that

$$\begin{cases} \ell_s = \lambda\mu\phi\varphi_s, & \ell_{x_j} = \lambda\mu\phi\varphi_{x_j}, & \ell_{sx_j} = \lambda\mu^2\phi\varphi_s\varphi_{x_j}, \\ \ell_{ss} = \lambda\mu^2\phi\varphi_s^2 + \lambda\mu\phi\varphi_{ss}, & \ell_{x_jx_k} = \lambda\mu^2\phi\varphi_{x_j}\varphi_{x_k} + \lambda\mu\phi\varphi_{x_jx_k}. \end{cases}$$
(2.13)

Recalling the definition of A, Ψ and Φ in (2.9) and (2.10), respectively, we have

$$\Phi = \lambda\phi O(\mu^2),$$
$$A = \sum_{j,k=1}^{n} h^{jk}\ell_{x_j}\ell_{x_k} + \ell_s^2 = \lambda^2\mu^2\phi^2\left(|\varphi_s|^2 + \sum_{j,k=1}^{n} h^{jk}\varphi_{x_j}\varphi_{x_k}\right). \tag{2.14}$$

Clearly, Ψ and Φ are chosen so that Ψ and A contain at most the first order partial derivatives of ϕ.

By the definition of c^{jk} and B in (2.12), and noting (2.10), (2.13) and (2.14), we obtain that

$$2\sum_{j,k=1}^{n} c^{jk}\,\mathrm{Re}\,(v_{x_j}\bar{v}_{x_k})$$

$$= 2\sum_{j,k=1}^{n}\left\{\sum_{j',k'=1}^{n}\left[2(h^{j'k}\ell_{x_{j'}})_{x_k}h^{jk'} - (h^{jk}h^{j'k'}\ell_{x_{j'}})_{x_k'}\right] - h^{jk}(\ell_{ss} + \Psi)\right\}\mathrm{Re}\,(v_{x_j}\bar{v}_{x_k})$$

$$= 4\lambda\mu^2\phi\left|\sum_{j,k=1}^{n} h^{jk}\varphi_{x_j}v_{x_k}\right|^2 + 2\left[\lambda\mu^2\phi\left(\sum_{j,k=1}^{n} h^{jk}\varphi_{x_j}\varphi_{x_k} + |\varphi_s|^2\right) + \lambda\phi O(\mu)\right]\sum_{j,k=1}^{n} h^{jk}v_{x_j}\bar{v}_{x_k} \tag{2.15}$$

and

$$B = 2(A\ell_s)_s + 2\sum_{j,k=1}^{n}(Ah^{jk}\ell_{x_j})_{x_k} + 2A\Psi - \Phi^2$$

$$= 2A_s\ell_s + 2\sum_{j,k=1}^{n} h^{jk}\ell_{x_j}A_{x_k} + 2A\left[\ell_{ss} + \sum_{j,k=1}^{n} h^{jk}\ell_{x_jx_k} + \Psi\right] + 2A\sum_{j,k=1}^{n} h^{jk}_{x_k}\ell_{x_j} - \Phi^2$$

$$= 2\lambda^3\mu^4\phi^3\left(|\varphi_s|^2 + \sum_{j,k=1}^{n} h^{jk}\varphi_{x_j}\varphi_{x_k}\right)^2 + \lambda^3\phi^3 O(\mu^3) + \lambda^2\phi^2 O(\mu^4). \tag{2.16}$$

It follows from (2.2), (2.15) and (2.16) that

$$2\sum_{j,k=1}^{n} c^{jk}\,\mathrm{Re}\,(v_{x_j}\bar{v}_{x_k}) + B|v|^2$$

$$\geq 4\lambda\mu^2\phi\left|\sum_{j,k=1}^{n} h^{jk}\varphi_{x_j}v_{x_k}\right|^2 + 2h_0\lambda\mu^2\phi\left(h_0|\nabla\varphi|^2 + |\varphi_s|^2\right)|\nabla v|^2$$

$$+ 2\lambda^3\mu^4\phi^3\left(h_0|\nabla\varphi|^2 + |\varphi_s|^2\right)^2|v|^2 - C\lambda\mu\phi\left(|\nabla v|^2 + \lambda^2\mu^2\phi^2|v|^2 + \lambda\mu^3\phi|v|^2\right). \tag{2.17}$$

Similarly, by (2.2), (2.10) and (2.13), we have

$$2\Big[\ell_{ss} - \sum_{j,k=1}^{n} (h^{jk}\ell_{x_j})_{x_k} - \Psi\Big]|v_s|^2$$
$$\geq 2\lambda\mu^2\phi\big(3|\varphi_s|^2 + h_0|\nabla\varphi|^2\big)|v_s|^2 - C\lambda\mu\phi|v_s|^2. \tag{2.18}$$

Further, by (2.13) and (2.14), recalling (2.10) for the definition of Ψ, we find that

$$8\sum_{j,k=1}^{n} h^{jk}\ell_{x_j s} \operatorname{Re}(v_{x_k}\bar{v}_s) - 2\sum_{j,k=1}^{n} h^{jk}\Psi_{x_k} \operatorname{Re}(v_{x_j}\bar{v}) - 2\Psi_s \operatorname{Re}(v_s\bar{v}) \tag{2.19}$$
$$\geq -4\lambda\mu^2\phi\Big|\sum_{j,k=1}^{n} h^{jk}\varphi_{x_j} v_{x_k}\Big|^2 - 4\lambda\mu^2\phi|\varphi_s|^2|v_s|^2 - C\big(\lambda^2\mu^3\phi^2|v|^2 + \mu^3|\nabla v|^2 + \mu^3|v_s|^2\big).$$

Combining (2.11) and (2.17)–(2.19), we conclude that there exist $c > 0$ and $\mu_0 > 1$, such that for every $\mu \geq \mu_0$, there exists $\lambda_0(\mu) = C\mu^2 > 0$ so that for all $\lambda \geq \lambda_0$, it holds that

$$4\theta^2\Big|z_{ss} + \sum_{j,k=1}^{n} (h^{jk}z_{x_j})_{x_k}\Big|^2 + M_s + \operatorname{div} V$$
$$\geq c\lambda\mu^2\phi\big(h_0|\nabla\varphi|^2 + |\varphi_s|^2\big)\big[|v_s|^2 + h_0|\nabla v|^2 + \lambda^2\mu^2\phi^2\big(h_0|\nabla\varphi|^2 + |\varphi_s|^2\big)|v|^2\big] \tag{2.20}$$
$$- C\lambda\mu\phi\big(|v_s|^2 + |\nabla v|^2 + \lambda^2\mu^2\phi^2|v|^2\big).$$

Step 3. End of the proof. Finally, recalling $v = \theta z$, it is easy to see that

$$\frac{1}{C}\theta^2\big(|\nabla z|^2 + |z_s|^2 + \lambda^2\mu^2\phi^2|\nabla\phi|^2|z|^2\big)$$
$$\leq |\nabla v|^2 + |v_s|^2 + \lambda^2\mu^2\phi^2\big(|\nabla\phi|^2 + |\phi_s|^2\big)|v|^2 \leq C\theta^2\big(|\nabla z|^2 + |z_s|^2 + \lambda^2\mu^2\phi^2|z|^2\big). \tag{2.21}$$

Integrating (2.20) in $(b_1, b_2) \times \Omega$, by (2.21) and noting that $z(b_1, \cdot) = z(b_2, \cdot) = z_s(b_1, \cdot) = z_s(b_2, \cdot) = 0$, we get the desired estimate (2.8) immediately.

Remark 2.3 In what follows, we apply Theorem 2.1 to establish some interpolation inequalities for elliptic equations in a cylindrical domain. In a similar way, one may derive interpolation inequalities for elliptic equations in general domains, which can be used to solve some inverse problems and unique continuation problems (e.g. [7]).

To present our second Carleman estimate, we need to introduce another weight function. For this purpose, for $\mu > 1$, we define

$$\tilde{\varphi}(r) = r \exp\Big(\int_0^r \frac{e^{-\mu t} - 1}{t}dt\Big), \qquad r > 0. \tag{2.22}$$

For $\lambda > 1$, put

$$\sigma(x) = |x|, \quad w(x) = \tilde{\varphi}(\sigma(x)), \quad \tilde{\ell}(x) = -\lambda \ln w(x), \quad \tilde{\theta}(x) = e^{\tilde{\ell}(x)}. \tag{2.23}$$

Remark 2.4 The reason for the above choice of $\tilde{\ell}$ is two fold. First, it is strongly pseudo-convex in $\Omega\setminus\{0\}$ (in the sense of [17, Definition 8.6.1] and [19, Definition 28.3.1]) for δ being small enough. Second, $w(x) = O(|x|)$ (as $|x| \to 0$), which is a key point in deriving the strong unique continuation property (SUCP for short) for solutions of elliptic equations (See Sect. 2.4 for more details).

By Lemma 2.1, one can also obtain the following Carleman estimate, which differs from Theorem 2.1 (Here, without loss of generality, we assume that $0 \in \Omega$ and $(h^{jk}(0))_{1\le j,k\le n}$ equals the identity matrix).

Theorem 2.2 *Let w be given in (2.23). Then, there exists $\mu_0 > 0$ such that for all $\mu > \mu_0$, one can find two constants $\lambda_0 = \lambda_0(\mu) > 0$ and $C > 0$ so that for all $z \in C_0^2(\Omega\setminus\{0\}; \mathbb{R})$ and $\lambda \ge \lambda_0$, it holds that*

$$\lambda^3 \mu \int_\Omega w^{-1-2\lambda} z^2 dx \le C \int_\Omega w^{2-2\lambda} \left| \sum_{j,k=1}^n (h^{jk} z_{x_j})_{x_k} \right|^2 dx. \tag{2.24}$$

Proof The proof is divided into three steps.

Step 1. For $\sigma = |x| > 0$, we define ϕ as follows:

$$\phi(\sigma) = \frac{\tilde{\varphi}(\sigma)}{\sigma \tilde{\varphi}'(\sigma)}. \tag{2.25}$$

Then, by (2.22), it is easy to see that

$$\phi(\sigma) = e^{\mu\sigma}, \quad \phi'(\sigma) = \mu\phi(\sigma). \tag{2.26}$$

Put

$$\mathscr{T} = \left(\sum_{j,k=1}^n h^{jk} \sigma_{x_j} \sigma_{x_k} \right)^{-1} = |x|^2 \left(\sum_{j,k=1}^n h^{jk} x_j x_k \right)^{-1}. \tag{2.27}$$

We shall use (2.3) by taking $m = n$, $(a^{jk})_{n\times n} = (h^{jk})_{n\times n}$, $\Psi = -\sum_{j,k=1}^n (h^{jk}\tilde{\ell}_{x_j})_{x_k}$, $\Phi = 0$ and $\theta(x) = \tilde{\theta}(x)$ (given by (2.23)). Multiplying (2.3) by $(\sigma\phi)^2\mathscr{T}$, we get that

$$(\sigma\phi)^2 \mathscr{T} \tilde{\theta}^2 \left| \sum_{j,k=1}^n (h^{jk} z_{x_j})_{x_k} \right|^2 + \text{div}\,(\sigma^2\phi^2\mathscr{T}V)$$

$$\ge (\sigma\phi)^2 \mathscr{T}|\tilde{I}_1|^2 + 2\sum_{j,k=1}^n \tilde{c}^{jk} v_{x_j} v_{x_k} + \nabla(\sigma^2\phi^2\mathscr{T}) \cdot V \tag{2.28}$$

$$- 2(\sigma\phi)^2 \mathscr{T} \sum_{j,k=1}^n h^{jk} \Psi_{x_j} v_{x_k} v + \tilde{B}v^2.$$

Here

$$\tilde{I}_1 = \sum_{j,k=1}^{n} (h^{jk} v_{x_j})_{x_k} + \tilde{A} v, \quad \tilde{A} = \sum_{j,k=1}^{n} h^{jk} \tilde{\ell}_{x_j} \tilde{\ell}_{x_k}, \tag{2.29}$$

and

$$\begin{cases} V = [V^1, \ldots, V^k, \ldots, V^n], \\ V^k = 2 \sum_{j=1}^{n} \Big[\sum_{j',k'=1}^{n} (2h^{jk'} h^{j'k} - h^{jk} h^{j'k'}) \tilde{\ell}_{x_j} v_{x_{j'}} v_{x_{k'}} - h^{jk} \Big(\Psi v_{x_j} v - A \tilde{\ell}_{x_j} |v|^2 \Big) \Big], \\ \tilde{c}^{jk} = (\sigma\phi)^2 \mathscr{T} \sum_{j',k'=1}^{n} \Big[2(h^{j'k} \tilde{\ell}_{x_{j'}})_{x_{k'}} h^{jk'} - h^{jk}_{x_{k'}} h^{j'k'} \tilde{\ell}_{x_{j'}} \Big], \\ \tilde{B} = 2(\sigma\phi)^2 \mathscr{T} \sum_{j,k=1}^{n} h^{jk} \tilde{\ell}_{x_j} \tilde{A}_{x_k}. \end{cases} \tag{2.30}$$

For simplicity, we put

$$\begin{cases} \mathscr{H}_1 = 2 \sum_{j,k,j',k'=1}^{n} (\sigma^2\phi^2 \mathscr{T})_{x_k} (2h^{jk'} h^{j'k} - h^{jk} h^{j'k'}) \tilde{\ell}_{x_j} v_{x_{j'}} v_{x_{k'}} + 2 \sum_{j,k=1}^{n} \tilde{c}^{jk} v_{x_j} v_{x_k}, \\ \mathscr{H}_2 = 2\tilde{A} \sum_{j,k=1}^{n} h^{jk} \tilde{\ell}_{x_j} (\sigma^2\phi^2 \mathscr{T})_{x_k} v^2 + \tilde{B} v^2, \\ \mathscr{H}_3 = -2 \sum_{j,k=1}^{n} h^{jk} \Psi v_{x_j} (\sigma^2\phi^2 \mathscr{T})_{x_k} v - 2(\sigma\phi)^2 \mathscr{T} \sum_{j,k=1}^{n} h^{jk} \Psi_{x_j} v_{x_k} v. \end{cases} \tag{2.31}$$

Then (2.28) can be rewritten as the following:

$$\begin{aligned} (\sigma\phi)^2 \mathscr{T} \tilde{\theta}^2 \Big| \sum_{j,k=1}^{n} (h^{jk} z_{x_j})_{x_k} \Big|^2 + \operatorname{div}(\sigma^2\phi^2 \mathscr{T} V) \\ \geq (\sigma\phi)^2 \mathscr{T} |\tilde{I}_1|^2 + \mathscr{H}_1 + \mathscr{H}_2 + \mathscr{H}_3. \end{aligned} \tag{2.32}$$

Step 2. Let us estimate $\mathscr{H}_1$, $\mathscr{H}_2$ and $\mathscr{H}_3$. By (2.23) and (2.25), it is easy to see that

$$\begin{aligned} \sigma_{x_j} &= \frac{x_j}{\sigma}, \quad \sigma_{x_j x_k} = \frac{\delta_{jk}}{\sigma} - \frac{1}{\sigma} \sigma_{x_j} \sigma_{x_k}, \\ w_{x_j} &= \frac{\tilde{\varphi}}{\sigma\phi} \sigma_{x_j}, \quad w_{x_j x_k} = \frac{\tilde{\varphi}(1 - \phi - \mu\sigma\phi)}{(\sigma\phi)^2} \sigma_{x_j} \sigma_{x_k} + \frac{\tilde{\varphi} \sigma_{x_j x_k}}{\sigma\phi}, \quad j, k = 1, 2, \ldots, n. \end{aligned} \tag{2.33}$$

Next, by (2.23) and (2.33), we have

$$\tilde{\ell}_{x_j} = -\lambda w^{-1} w_{x_j} = -\lambda (\sigma\phi)^{-1} \sigma_{x_j}, \quad \forall j = 1, 2, \ldots, n, \tag{2.34}$$

and

$$
\begin{aligned}
\tilde{\ell}_{x_j x_k} &= \lambda w^{-2} (w_{x_j} w_{x_k} - w w_{x_j x_k}) \\
&= \lambda (\sigma \phi)^{-2} \big[(2\phi + \mu \sigma \phi) \sigma_{x_j} \sigma_{x_k} - \delta_{jk} \phi \big], \quad \forall j, k = 1, 2, \ldots, n.
\end{aligned} \tag{2.35}
$$

Further, by (2.27) and recalling that $(h^{jk}(0))_{1 \le j, k \le n}$ is assumed to be the identity matrix, we obtain that

$$
\begin{aligned}
\mathcal{T}_{x_j} &= \Big(\sum_{j,k=1}^{n} h^{jk} x_j x_k \Big)^{-2} \Big(2 x_j \sum_{j,k=1}^{n} h^{jk} x_j x_k - 2|x|^2 \sum_{k=1}^{n} h^{jk} x_k - |x|^2 \sum_{j,k=1}^{n} h^{jk}_{x_j} x_j x_k \Big) \\
&= \Big(\sum_{j,k=1}^{n} h^{jk} x_j x_k \Big)^{-2} \Big[2 x_j \sum_{j,k=1}^{n} h^{jk}(0) x_j x_k + 2 x_j \sum_{j,k=1}^{n} \big(h^{jk}(x) - h^{jk}(0) \big) x_j x_k \\
&\quad - 2|x|^2 \sum_{k=1}^{n} h^{jk}(0) x_k + 2|x|^2 \sum_{k=1}^{n} \big(h^{jk}(x) - h^{jk}(0) \big) x_k - |x|^2 \sum_{j,k=1}^{n} h^{jk}_{x_j} x_j x_k \Big] \\
&= \Big(\sum_{j,k=1}^{n} h^{jk} x_j x_k \Big)^{-2} \Big[2 x_j \sum_{j,k=1}^{n} \big(h^{jk}(x) - h^{jk}(0) \big) x_j x_k \\
&\quad - 2|x|^2 \sum_{k=1}^{n} \big(h^{jk}(x) - h^{jk}(0) \big) x_k - |x|^2 \sum_{j,k=1}^{n} h^{jk}_{x_j} x_j x_k \Big].
\end{aligned} \tag{2.36}
$$

Since $h^{jk}(\cdot) \in C^1(\overline{\Omega}; \mathbb{R})$ for $j, k = 1, \ldots, n$, for any $x \in \overline{\Omega}$, there is a $C = C(h^{jk})$ such that

$$
\max_{1 \le j, k \le n} \{ |h^{jk}(x) - h^{jk}(0)| \} \le C|x|, \quad \forall x \in \overline{\Omega}.
$$

Hence,

$$
\Big| 2 x_j \sum_{j,k=1}^{n} \big(h^{jk}(x) - h^{jk}(0) \big) x_j x_k \Big| + \Big| 2|x|^2 \sum_{k=1}^{n} \big(h^{jk}(x) - h^{jk}(0) \big) x_k \Big| \le C|x|^4.
$$

This implies that

$$
\mathcal{T}_{x_j} = O(1), \quad j = 1, \ldots, n, \text{ as } |x| \to 0. \tag{2.37}
$$

By (2.31) and (2.33)–(2.37), we have

$$
\begin{aligned}
\mathcal{H}_1 &= 4 \sum_{j,k,j',k'=1}^{n} (\sigma^2 \phi^2 \mathcal{T} \tilde{\ell}_{x_{j'}})_{x_{k'}} h^{j'k} h^{jk'} v_{x_j} v_{x_k} \\
&\quad - 2 \sum_{j,k,j',k'=1}^{n} (\sigma^2 \phi^2 \mathcal{T})_{x_{k'}} h^{jk} h^{j'k'} \tilde{\ell}_{x_{j'}} v_{x_j} v_{x_k} + \sigma \phi O(\lambda) |\nabla v|^2,
\end{aligned}
$$

$$
= -4\lambda \sum_{j,k,j',k'=1}^{n} (\sigma\phi\sigma_{x_{j'}})_{x_{k'}} \mathcal{T} h^{j'k} h^{jk'} v_{x_j} v_{x_k}
$$

$$
+ 4\lambda(1 + \mu\sigma)\phi \sum_{j,k=1}^{n} h^{jk} v_{x_j} v_{x_k} + \sigma\phi O(\lambda)|\nabla v|^2,
$$

$$
= -4\lambda\mu\sigma\phi\mathcal{T} \left| \sum_{j,k=1}^{n} h^{jk}\sigma_{x_j} v_{x_k} \right|^2 + 4\lambda\mu\sigma\phi \sum_{j,k=1}^{n} h^{jk} v_{x_j} v_{x_k}
$$

$$
+ 4\lambda\phi \sum_{j,k=1}^{n} \left[h^{jk} - \mathcal{T} \sum_{j',k'=1}^{n} h^{j'k} h^{jk'} \delta_{j'k'} \right] v_{x_j} v_{x_k} + \sigma\phi O(\lambda)|\nabla v|^2
$$

$$
= -4\lambda\mu\sigma\phi\mathcal{T} \left| \sum_{j,k=1}^{n} h^{jk}\sigma_{x_j} v_{x_k} \right|^2 + 4\lambda\mu\sigma\phi \sum_{j,k=1}^{n} h^{jk} v_{x_j} v_{x_k} + \sigma\phi O(\lambda)|\nabla v|^2.
$$

$$(2.38)$$

Further, recalling (2.30) for $\tilde{B}$, by (2.31), (2.29) and (2.27), it is easy to see that

$$
\mathscr{H}_2 = 2 \sum_{j,k=1}^{n} h^{jk}\tilde{\ell}_{x_j} (\tilde{A}\sigma^2\phi^2 \mathcal{T})_{x_k} v^2 = 0. \tag{2.39}
$$

Recall that $\Psi = - \sum_{j,k=1}^{n} (h^{jk}\tilde{\ell}_{x_j})_{x_k}$. Hence, by (2.34)–(2.35), we have that

$$
\sigma^2\phi^2 \mathcal{T}\Psi = -\sigma^2\phi^2 \mathcal{T} \sum_{j,k=1}^{n} (h^{jk}\tilde{\ell}_{x_j})_{x_k} = -\lambda\mu\sigma\phi + O(\lambda)\phi. \tag{2.40}
$$

By (2.31) and (2.33)–(2.35), we have

$$
\mathscr{H}_3 = - \sum_{j,k=1}^{n} h^{jk}(v^2)_{x_j} (\Psi\sigma^2\phi^2 \mathcal{T})_{x_k}
$$

$$
= - \sum_{j,k=1}^{n} \left[h^{jk}(v^2)_{x_j} \Psi\sigma^2\phi^2 \mathcal{T} \right]_{x_k} + \Psi\sigma^2\phi^2 \mathcal{T} \sum_{j,k=1}^{n} \left[h^{jk}(v^2)_{x_j} \right]_{x_k}
$$

$$
= - \sum_{j,k=1}^{n} \left[h^{jk}(v^2)_{x_j} \Psi\sigma^2\phi^2 \mathcal{T} \right]_{x_k} - \lambda\mu\sigma\phi \sum_{j,k=1}^{n} \left[h^{jk}(v^2)_{x_j} \right]_{x_k}
$$

$$
+ O(\lambda)\left[\phi \sum_{j,k=1}^{n} h^{jk}(v^2)_{x_j} \right]_{x_k} - O(\lambda)\left(\sum_{j,k=1}^{n} h^{jk}\phi_{x_k} v^2 \right)_{x_j} + O(\lambda)v^2 \sum_{j,k=1}^{n} (h^{jk}\phi_{x_j})_{x_k}.
$$

$$(2.41)$$

By (2.29), it follows that

$$
\begin{aligned}
-\lambda\mu\sigma\phi & \sum_{j,k=1}^{n} \left[h^{jk}(v^2)_{x_j} \right]_{x_k} \\
&= -2\lambda\mu\sigma\phi v \sum_{j,k=1}^{n} (h^{jk}v_{x_j})_{x_k} - 2\lambda\mu\sigma\phi \sum_{j,k=1}^{n} h^{jk}v_{x_j}v_{x_k} \\
&= -2\lambda\mu\sigma\phi v\left(\tilde{I}_1 - \sum_{j,k=1}^{n} h^{jk}\tilde{\ell}_{x_j}\tilde{\ell}_{x_k}v \right) - 2\lambda\mu\sigma\phi \sum_{j,k=1}^{n} h^{jk}v_{x_j}v_{x_k} \\
&= 2\lambda^3\mu(\sigma\phi)^{-1} \sum_{j,k=1}^{n} h^{jk}\sigma_{x_j}\sigma_{x_k}v^2 - 2\lambda\mu\sigma\phi \sum_{j,k=1}^{n} h^{jk}v_{x_j}v_{x_k} - 2\lambda\mu\sigma\phi v\tilde{I}_1.
\end{aligned}
\tag{2.42}
$$

Now, we integrate (2.28) on Ω. By (2.38)–(2.39), (2.41)–(2.42) and (2.2), for all $z \in C_0^2(\Omega\backslash\{0\}; \mathbb{R})$, we have that

$$
\begin{aligned}
\int_{\Omega} (\sigma\phi)^2 |\tilde{I}_1|^2 dx &+ \lambda\mu h_0 \int_{\Omega} \sigma\phi|\nabla v|^2 dx + \lambda^3\mu \int_{\Omega} (\sigma\phi)^{-1}v^2 dx \\
&\leq C \int_{\Omega} (\sigma\phi)^2\theta^2 \left| \sum_{j,k=1}^{n} (h^{jk}z_{x_j})_{x_k} \right|^2 dx + C \int_{\Omega} \mathscr{H} dx,
\end{aligned}
\tag{2.43}
$$

where

$$
\begin{aligned}
\mathscr{H} = \lambda\mu\sigma\phi & \left| \sum_{j,k=1}^{n} h^{jk}\sigma_{x_j}v_{x_k} \right|^2 + \lambda\sigma\phi|\nabla v|^2 \\
&+ \lambda \left| \sum_{j,k=1}^{n} (h^{jk}\phi_{x_j})_{x_k} \right| v^2 + \lambda\mu\sigma\phi|v||I_1|.
\end{aligned}
\tag{2.44}
$$

Step 3. Let us estimate $\mathscr{H}$. By $\phi(\sigma) = e^{\mu\sigma}$, we have

$$
\sum_{j,k=1}^{n} (h^{jk}\phi_{x_j})_{x_k} = \mu^2\phi \sum_{j,k=1}^{n} h^{jk}\sigma_{x_j}\sigma_{x_k} + \mu\phi \sum_{j,k=1}^{n} (h^{jk}\sigma_{x_j})_{x_k}.
\tag{2.45}
$$

On the other hand, by (2.22) and (2.33), it follows that for a constant $C_1 > 0$,

$$
\frac{\sigma}{C_1} \leq w \leq C_1\sigma, \qquad \frac{1}{C_1} \leq |\nabla w| \leq C_1.
\tag{2.46}
$$

Hence, by (2.44)–(2.46), we have

$$\mathcal{H} \le C\left(\lambda\mu w \left| \sum_{j,k=1}^{n} h^{jk}\sigma_{x_j}v_{x_k}\right|^2 + \lambda w|\nabla v|^2 + \lambda\mu w^{-1}v^2 + \lambda^2\mu^2 v^2\right)$$
$$+ \frac{1}{2}(\sigma\phi)^2|\tilde{I}_1|^2. \tag{2.47}$$

By (2.43), (2.46) and (2.47), we conclude that there is a constant $\mu_0 > 0$ such that for all $\mu > \mu_0$, one can find $\lambda_1 = \lambda_1(\mu) > 0$ so that for all $z \in C_0^2(\Omega\setminus\{0\}; \mathbb{R})$ and $\lambda \ge \lambda_1$, it holds that

$$\lambda\mu \int_\Omega w\left(h_0|\nabla v|^2 + \lambda^2 w^{-2}v^2\right)dx + \int_\Omega w^2|\tilde{I}_1|^2 dx$$
$$\le C\int_\Omega w^{2-2\lambda}\left|\sum_{j,k=1}^{n}(h^{jk}z_{x_j})_{x_k}\right|^2 dx + C\lambda\mu\int_\Omega w\left|\sum_{j,k=1}^{n}h^{jk}\sigma_{x_j}v_{x_k}\right|^2 dx. \tag{2.48}$$

Recall (2.30) for $\tilde{I}_1$, by (2.34), it is easy to check that

$$\tilde{\theta}\sum_{j,k=1}^{n}(h^{jk}z_{x_j})_{x_k} - \tilde{I}_1 = -2\sum_{j,k=1}^{n}h^{jk}\tilde{\ell}_{x_j}v_{x_k} - \sum_{j,k=1}^{n}(h^{jk}\tilde{\ell}_{x_j})_{x_k}v$$
$$= 2\lambda(\sigma\phi)^{-1}\sum_{j,k=1}^{n}h^{jk}\sigma_{x_j}v_{x_k} - \sum_{j,k=1}^{n}(h^{jk}\tilde{\ell}_{x_j})_{x_k}v. \tag{2.49}$$

By (2.35), we find $\left|\sum_{j,k=1}^{n}(h^{jk}\tilde{\ell}_{x_j})_{x_k}\right| \le C\lambda\mu w^{-2}$. This, together with (2.49), implies that

$$\lambda\mu w\left|\sum_{j,k=1}^{n}h^{jk}\sigma_{x_j}v_{x_k}\right|^2$$
$$= \lambda\mu w\frac{\sigma^2\phi^2}{4\lambda^2}\left|\tilde{\theta}\sum_{j,k=1}^{n}(h^{jk}z_{x_j})_{x_k} - \tilde{I}_1 + \sum_{j,k=1}^{n}(h^{jk}\tilde{\ell}_{x_j})_{x_k}v\right|^2 \tag{2.50}$$
$$\le \frac{C\mu\sigma^2\phi^2 w}{\lambda}\left[\tilde{\theta}^2\left|\sum_{j,k=1}^{n}(h^{jk}z_{x_j})_{x_k}\right|^2 + |\tilde{I}_1|^2\right] + C\lambda\mu^3 w^{-1}v^2.$$

Finally, combining (2.48) and (2.50), noting that $v = w^{-\lambda}z$, we conclude that there exists $\lambda_0 > 0$ so that (2.24) holds for any $\lambda > \lambda_0(\mu)$.

Remark 2.5 Carleman estimate in the form of Theorem 2.2 was first established in [18]. Here, we provide a slightly different proof.

2.2 Observability Estimate for Finite Sums of Eigenfunctions of Elliptic Operators

As an application of Theorem 2.1, in this section, we prove an observability estimate for finite sums of eigenfunctions of some second order elliptic operators.

Define an unbounded operator $\mathscr{A}$ on $L^2(\Omega)$ as follows:

$$\begin{cases} D(\mathscr{A}) = H^2(\Omega) \cap H_0^1(\Omega), \\ \mathscr{A}u = - \displaystyle\sum_{j,k=1}^{n} (h^{jk}u_{x_j})_{x_k}, \quad \forall\, u \in D(\mathscr{A}). \end{cases}$$

Let $\{\lambda_k\}_{k=1}^{\infty}$ be the eigenvalues of $\mathscr{A}$, and $\{e_k\}_{k=1}^{\infty}$ be the corresponding eigenfunctions satisfying $|e_k|_{L^2(\Omega)} = 1$, $k = 1, 2, 3 \ldots$. It is well known that $0 < \lambda_1 < \lambda_2 \leq \ldots$, and $\{e_k\}_{k=1}^{\infty}$ constitutes an orthonormal basis of $L^2(\Omega)$.

We have the following observability estimate for finite sums of the eigenfunctions of $\mathscr{A}$:

Theorem 2.3 *There exist two positive constants C_1 and C_2 such that*

$$\sum_{\lambda_k \leq r} |a_k|^2 \leq C_1 e^{C_2 \sqrt{r}} \int_{\omega} \Big| \sum_{\lambda_k \leq r} a_k e_k(x) \Big|^2 dx \tag{2.51}$$

holds for every $r \in (0, \infty)$ and any choice of coefficients $\{a_k\}_{\lambda_k \leq r} \subset \mathbb{C}$.

Remark 2.6 The inequality (2.51) is sharp in the sense that the $\sqrt{r}$ in (2.51) cannot be replaced by r^α for any $\alpha < \frac{1}{2}$ (See [22] for more details).

Remark 2.7 The observability estimate for a single eigenfunction of an elliptic operator was first obtained in [9] to estimate the Hausdorff dimension of the nodal sets of eigenfunctions. The result in the form of (2.51) was first proved in [34] (an equivalent form was obtained in [32]) when Γ is C^∞. The result was generalized for the case that h^{jk} is piecewise C^∞ in [29]. The case that Γ is C^2 was established in [40]. In [40], it was also shown that (2.51) holds for eigenfunctions of elliptic operators with homogeneous Robin boundary condition; while in [43], it was proved that (2.51) holds for eigenfunctions of elliptic operators with homogeneous equivalued surface boundary condition. Recently, (2.51) was generalized to bi-Lapliacian operator in [28].

Remark 2.8 The inequality (2.51) has several applications in PDE control problems (see [37, 39, 41–44, 49] and the references therein). Further, it can be applied to study the nodal sets for eigenfunctions of Laplacian operators ([22]).

In order to prove Theorem 2.3, people introduce the following elliptic equation (in which the unknown $u \equiv u(s, x)$):

$$\begin{cases} u_{ss} + \displaystyle\sum_{j,k=1}^{n} (h^{jk} u_{x_j})_{x_k} = 0 & \text{in } (0, 4) \times \Omega, \\ u = 0 & \text{on } (0, 4) \times \Gamma. \end{cases} \tag{2.52}$$

The following interpolation inequality holds for solutions to (2.52):

Lemma 2.2 *There exist two constants $\kappa \in (0, 1)$ and $C > 0$ such that any solution $u \in H^2((0, 4) \times \Omega; \mathbb{C})$ to (2.52) with $u = 0$ on $\{0\} \times \Omega$ satisfies that*

$$|u|_{L^2((1,3)\times\Omega)} \leq C |u_s(0)|^{\kappa}_{L^2(\omega)} |u|^{1-\kappa}_{H^1((0,4)\times\Omega)}. \tag{2.53}$$

Remark 2.9 Similarly, one can prove the inequality (2.53) for solutions to elliptic equations in $(0, T) \times \Omega$ for any $T > 0$. Here, to simplify notation, we choose $T = 4$.

Lemma 2.2 follows immediately from Lemma 2.3 (Interpolation inequality I) and Lemma 2.5 (Interpolation inequality II), which are presented in the next two subsections. Based on Lemma 2.2, we may prove Theorem 2.3 as follows:

Proof of Theorem 2.3. Set

$$u(s, x) = \sum_{\lambda_k \leq r} \frac{\text{sh}(s\sqrt{\lambda_k})}{\sqrt{\lambda_k}} a_k e_k(x).$$

Then u is a solution to (2.52) and $u = 0$ on $\{0\} \times \Omega$. For the left hand side of (2.53), we have

$$|u|^2_{L^2((1,3)\times\Omega)} = \int_1^3 \int_\Omega \Big| \sum_{\lambda_k \leq r} \frac{\text{sh}(s\sqrt{\lambda_k})}{\sqrt{\lambda_k}} a_k e_k \Big|^2 dx ds$$

$$= \sum_{\lambda_k \leq r} |a_k|^2 \int_1^3 \int_\Omega \Big| \frac{\text{sh}(s\sqrt{\lambda_k})}{\sqrt{\lambda_k}} \Big|^2 dx ds \geq \sum_{\lambda_k \leq r} |a_k|^2 \int_1^3 s^2 ds = \frac{26}{3} \sum_{\lambda_k \leq r} |a_k|^2.$$

On the other hand, $u_s(0, x) = \displaystyle\sum_{\lambda_j \leq r} a_j e_j$ and

$$|u|^2_{H^1((0,4)\times\Omega)} \leq C e^{8\sqrt{r}} (1 + r) \sum_{\lambda_k \leq r} |a_k|^2 \leq C e^{9\sqrt{r}} \sum_{\lambda_k \leq r} |a_k|^2.$$

Consequently, by Lemma 2.2, we have

$$\sum_{\lambda_k \leq r} |a_k|^2 \leq C \Big(\int_\omega \Big| \sum_{\lambda_k \leq r} a_k e_k \Big|^2 dx \Big)^\kappa \Big(e^{9\sqrt{r}} \sum_{\lambda_k \leq r} |a_k|^2 \Big)^{1-\kappa},$$

which implies that

$$\sum_{\lambda_k \leq r} |a_k|^2 \leq C_1 e^{C_2 \sqrt{r}} \int_\omega \left| \sum_{\lambda_k \leq r} a_k e_k \right|^2 dx.$$

This completes the proof of Theorem 2.3.

2.2.1 Interpolation Inequality I

We now show the following interpolation inequality:

Lemma 2.3 *Let $0 < \gamma < 1$. Then there exists a constant $C > 0$ such that for all $\varepsilon > 0$ and every solution $u \in H^2((0,4) \times \Omega; \mathbb{C})$ to (2.52), it holds that*

$$|u|_{H^1((1,3)\times\Omega)} \leq Ce^{C/\varepsilon} |u|_{L^2((\gamma,4-\gamma)\times\omega)} + Ce^{-2/\varepsilon} |u|_{H^1((0,4)\times\Omega)}. \tag{2.54}$$

Remark 2.10 By a simple change of time variable t to $t + 2$, in order to prove Lemma 2.3, it suffices to prove that

$$|u|_{H^1((-1,1)\times\Omega)} \leq Ce^{C/\varepsilon} \left(|g|_{L^2((-2,2)\times\Omega)} + |u|_{L^2((-2+\gamma,2-\gamma)\times\omega)} \right) + Ce^{-2/\varepsilon} |u|_{H^1((-2,2)\times\Omega)}, \tag{2.55}$$

where $g \in L^2((-2,2) \times \Omega; \mathbb{C})$ and u solves the following elliptic equation:

$$\begin{cases} u_{ss} + \displaystyle\sum_{j,k=1}^n (h^{jk} u_{x_j})_{x_k} = g & \text{in } (-2,2) \times \Omega, \\ u = 0 & \text{on } (-2,2) \times \Gamma. \end{cases} \tag{2.56}$$

Remark 2.11 By Lemma 2.3, one can establish a logarithmic decay result for solutions to hyperbolic equations with suitable damping (See Sect. 2.3 for the detail).

Remark 2.12 In case of $g = 0$, according to [40], the inequality (2.54) has another form, that is, there exist two constants $\kappa \in (0,1)$ and $C > 0$ such that every solution $u \in H^1((0,4) \times \Omega; \mathbb{C})$ to (2.52) satisfies that

$$|u|_{H^1((1,3)\times\Omega)} \leq C|u|_{L^2((\gamma,4-\gamma)\times\omega)}^\kappa |u|_{H^1((0,4)\times\Omega)}^{1-\kappa}. \tag{2.57}$$

In fact, by (2.54), there exists $\hat{\kappa} > 0$ such that for any $\varepsilon > 0$,

$$|u|_{H^1((1,3)\times\Omega)} \leq \varepsilon^{-\hat{\kappa}} |u|_{L^2((\gamma,4-\gamma)\times\omega)} + C\varepsilon |u|_{H^1((0,4)\times\Omega)}. \tag{2.58}$$

Putting $\kappa = \dfrac{1}{1+\hat{\kappa}}$ and $\varepsilon = \left(\dfrac{|u|_{L^2((\gamma,4-\gamma)\times\omega)}}{|u|_{H^1((0,4)\times\Omega)}} \right)^\kappa$ in (2.58), we obtain (2.57) immediately.

Recall that ω_0 is a nonempty open subset such that $\overline{\omega}_0 \subset \omega \subset \Omega$. The following known result is a basis to prove Lemma 2.3.

Lemma 2.4 ([15, Lemma 1.1]) *There exists a function $\psi \in C^2(\overline{\Omega}; \mathbb{R})$ such that $\psi > 0$ in Ω, $\psi = 0$ on Γ and*

$$|\nabla \psi| > 0, \quad \forall x \in \overline{\Omega} \setminus \omega_0. \tag{2.59}$$

We are now in a position to give a proof of Lemma 2.3.

Proof of Lemma 2.3. As mentioned in Remark 2.10, we only need to prove the inequality (2.55). The proof is divided into three steps.

Step 1. Note that $0 < \gamma < 1$, then $1 < b \overset{\triangle}{=} 2 - \gamma < 2$. Since there is no boundary condition for u at $s = \pm 2$ in (2.56), we need to introduce a cut-off function $q = q(\cdot) \in C_0^\infty(-b, b) \subset C_0^\infty(\mathbb{R})$ such that

$$\begin{cases} 0 \le q(s) \le 1, \ |s| < b, \\ q(s) = 1, \qquad |s| \le b_0, \end{cases} \tag{2.60}$$

where b and b_0 (satisfying $1 < b_0 < b < 2$) are given as follows:

$$b \overset{\triangle}{=} \sqrt{1 + \frac{1}{\mu} \ln(2 + e^\mu)}, \quad b_0 \overset{\triangle}{=} \sqrt{b^2 - \frac{1}{\mu} \ln\left(\frac{1 + e^\mu}{e^\mu}\right)}. \tag{2.61}$$

In (2.61), $\mu > \ln 2$ is the parameter that appeared in Theorem 2.1 and will be chosen large enough.

Put $z = qu$. Since q does not depend on x, it follows from (2.56) that

$$\begin{cases} z_{ss} + \sum_{j,k=1}^n \left(h^{jk} z_{x_j}\right)_{x_k} = q_{ss}u + 2q_s u_s + qg \quad \text{in } (-2, 2) \times \Omega, \\ z = 0 \qquad \qquad \qquad \qquad \qquad \text{on } (-2, 2) \times \Gamma. \end{cases} \tag{2.62}$$

Let us choose the weight function appeared in Lemma 2.1 as follows:

$$\varphi = \varphi(s, x) \overset{\triangle}{=} \frac{\psi(x)}{\|\psi\|_{L^\infty(\Omega)}} + b^2 - s^2, \quad \phi = e^{\mu\varphi}, \quad \theta = e^\ell = e^{\lambda\phi}, \tag{2.63}$$

where ψ is given by Lemma 2.4. By (2.59) and (2.63), we find that

$$h \overset{\triangle}{=} |\nabla\varphi| = \frac{1}{\|\psi\|_{L^\infty(\Omega)}} |\nabla\psi| > 0, \quad \text{in } \overline{\Omega} \setminus \omega_0, \tag{2.64}$$

and

$$\begin{cases} \phi(s, \cdot) \ge 2 + e^\mu, \quad \text{for all } s \text{ satisfying } |s| \le 1, \\ \phi(s, \cdot) \le 1 + e^\mu, \quad \text{for all } s \text{ satisfying } b_0 \le |s| \le b. \end{cases} \tag{2.65}$$

Step 2. Noting that $v = \theta z$ and $z = 0$ on $(-2, 2) \times \Gamma$, we have

$$v_s = 0, \quad v_{x_j} = \frac{\partial v}{\partial \nu} v^j, \quad \bar{v}_{x_j} = \frac{\partial \bar{v}}{\partial \nu} v^j, \quad j = 1, \ldots, n \quad \text{on } (-2, 2) \times \Gamma. \tag{2.66}$$

Since $\psi > 0$ in Ω and $\psi = 0$ on $(-2, 2) \times \Gamma$, we see that

$$\frac{\partial \varphi}{\partial \nu} \leq 0 \quad \text{on } (-2, 2) \times \Gamma. \tag{2.67}$$

Let us take $b_1 = -b$ and $b_2 = b$ in Theorem 2.1. It follows from (2.13), (2.63), (2.66) and (2.67) that

$$\int_{-b}^{b} \int_{\Gamma} \left(\sum_{j,k=1}^{n} h^{jk} \ell_{x_j} v^k \sum_{j',k'=1}^{n} h^{j'k'} v^{j'} v^{k'} \right) \left| \frac{\partial v}{\partial \nu} \right|^2 d\Gamma ds$$

$$= 2\lambda\mu \int_{-b}^{b} \int_{\Gamma} \phi \left(\sum_{j,k=1}^{n} h^{jk} v^j v^k \right)^2 \left| \frac{\partial v}{\partial \nu} \right|^2 \frac{\partial \varphi}{\partial \nu} d\Gamma ds \leq 0. \tag{2.68}$$

By (2.8), (2.62), (2.64) and (2.68), we see that

$$\int_{-b}^{b} \int_{\Omega} \lambda\mu^2 h^2 \theta^2 \left[\phi \left(|\nabla z|^2 + |z_s|^2 + \lambda^2 \mu^2 \phi^2 h^2 |z|^2 \right) \right] dxds$$

$$\leq C \left[\int_{-b}^{b} \int_{\Omega} \theta^2 |q_{ss} u + 2 q_s u_s + qg|^2 dxds \right. \tag{2.69}$$

$$\left. + \lambda\mu \int_{-b}^{b} \int_{\Omega} \phi \theta^2 \left(\lambda^2 \mu^2 \phi^2 |z|^2 + |z_s|^2 + |\nabla z|^2 \right) dxds \right].$$

On the other hand, by (2.64) and Lemma 2.4, we have

The left side of (2.69)

$$\geq c\lambda\mu^2 \int_{-b}^{b} \int_{\Omega \setminus \omega_0} \theta^2 \phi \left(|z_s|^2 + |\nabla z|^2 + \lambda^2 \mu^2 \phi^2 |z|^2 \right) dxds \tag{2.70}$$

$$- C\lambda\mu^2 \int_{-b}^{b} \int_{\omega_0} \theta^2 \phi \left(|z_s|^2 + |\nabla z|^2 + \lambda^2 \mu^2 \phi^2 |z|^2 \right) dxds,$$

where c is a positive constant.

Combining (2.69) and (2.70), taking $\lambda \geq 1 + C$, we end up with

$$\lambda\mu^2 \int_{-b}^{b} \int_{\Omega} \theta^2 \phi \left(|\nabla z|^2 + |z_s|^2 + \lambda^2 \mu^2 \phi^2 |z|^2 \right) dxds$$

$$\leq C \int_{-b}^{b} \int_{\Omega} \theta^2 |q_{ss} u + 2 q_s u_s + qg|^2 dxds \tag{2.71}$$

$$+ Ce^{C\lambda} \int_{-b}^{b} \int_{\omega_0} \left(|z|^2 + |\nabla z|^2 + |z_s|^2 \right) dxds.$$

Step 3. Now, we choose a cut-off function $\zeta \in C_0^2(\omega; [0, 1])$ such that $\zeta(\cdot) = 1$ on ω_0. Multiplying (2.62) by $\zeta^2 \bar{z}$, we obtain that

$$
\begin{aligned}
\zeta^2 \bar{z} \Big(z_{ss} + \sum_{j,k=1}^n (h^{jk}(z_{x_j})_{x_k}) \Big) \\
= (\zeta^2 \bar{z} z_s)_s - \zeta^2 |z_s|^2 - \sum_{j,k=1}^n (\zeta^2 \bar{z} h^{jk} z_{x_j})_{x_k} - \zeta^2 \sum_{j,k=1}^n h^{jk} z_{x_j} \bar{z}_{x_k} \\
- 2\zeta \bar{z} \sum_{j,k=1}^n h^{jk} z_{x_j} \zeta_{x_k}.
\end{aligned}
\tag{2.72}
$$

Integrating (2.72) in $(-b, b) \times \Omega$, noting that $z(-b, \cdot) = z(b, \cdot) \equiv 0$ and $z = 0$ on the boundary, we arrive at

$$
\begin{aligned}
& \int_{-b}^b \int_\Omega \zeta^2 \Big(|z_s|^2 + \sum_{j,k=1}^n h^{jk} z_{x_j} \bar{z}_{x_k} \Big) dx ds \\
& \leq \int_{-b}^b \int_\Omega \zeta^2 \bar{z} (q_{ss} u + 2q_s u_s + qg) dx ds \\
& \quad + \Big(\int_{-b}^b \int_\Omega \zeta^2 \sum_{j,k=1}^n h^{jk} z_{x_j} \bar{z}_{x_k} dx ds \Big)^{\frac{1}{2}} \Big(\int_{-b}^b \int_\Omega 4|z|^2 \sum_{j,k=1}^n h^{jk} \zeta_{x_j} \zeta_{x_k} dx ds \Big)^{\frac{1}{2}} \\
& \leq C \Big(\int_{-b}^b \int_\Omega \zeta^2 |q_{ss} u + 2q_s u_s + qg|^2 dx ds + \int_{-b}^b \int_\Omega \zeta^2 |z|^2 dx ds \Big) \\
& \quad + \frac{1}{2} \int_{-b}^b \int_\Omega \zeta^2 \sum_{j,k=1}^n h^{jk} z_{x_j} \bar{z}_{x_k} dx ds + 8 \int_{-b}^b \int_\Omega |z|^2 \sum_{j,k=1}^n h^{jk} \zeta_{x_j} \zeta_{x_k} dx ds.
\end{aligned}
\tag{2.73}
$$

By (2.2), (2.71) and (2.73), we end up with

$$
\begin{aligned}
& \lambda \mu^2 \int_{-b}^b \int_\Omega \theta^2 \phi \big(|\nabla z|^2 + |z_s|^2 + \lambda^2 \mu^2 \phi^2 |z|^2 \big) dx ds \\
& \leq C \Big(\int_{-b}^b \int_\Omega \theta^2 |q_{ss} u + 2q_s u_s + qg|^2 dx ds + e^{C\lambda} \int_{-b}^b \int_{\omega_0} |z|^2 dx ds \Big).
\end{aligned}
\tag{2.74}
$$

Let $c_0 = 2 + e^\mu > 1$, and recall (2.61) for $b_0 \in (1, b)$ and $b = 2 - \gamma \in (1, 2)$. Fixing μ in (2.74), using (2.60) and (2.65), noting that $z = qu$, we conclude that

$$
\begin{aligned}
& \lambda e^{2\lambda c_0} \int_{-1}^1 \int_\Omega \big(|\nabla u|^2 + |u_s|^2 + |u|^2 \big) dx ds \\
& \leq C e^{C\lambda} \Big(\int_{-2}^2 \int_\Omega |g|^2 dx ds + \int_{-2+\gamma}^{2-\gamma} \int_\omega |u|^2 dx ds \Big) \\
& \quad + C e^{2\lambda(c_0-1)} \int_{(-b,-b_0) \cup (b_0,b)} \int_\Omega \big(|u|^2 + |u_s|^2 \big) dx ds.
\end{aligned}
\tag{2.75}
$$

From (2.75), one can find an $\varepsilon_2 > 0$ such that the desired inequality (2.55) holds for $\varepsilon \in (0, \varepsilon_2]$, which, in turn, implies that it holds for all $\varepsilon > 0$. This completes the proof of the inequality (2.55).

2.2.2 Interpolation Inequality II

Lemma 2.5 *Let $0 < \gamma < 1$. Then there exist two constants $\delta \in (0, 1)$ and $C > 0$ such that every solution $u \in H^2((0, 4) \times \Omega; \mathbb{C})$ to (2.52) satisfies that*

$$|u|_{H^1((\gamma, 4-\gamma) \times \omega)} \leq C\big(|u(0)|_{L^2(\omega)} + |u_s(0)|_{L^2(\omega)} + |\nabla u(0)|_{L^2(\omega)}\big)^\delta |u|_{H^1((0,4) \times \Omega)}^{1-\delta}.$$

Proof We divide the proof of Lemma 2.5 into three steps.

Step 1. Choice of the weight functions. Let $\omega_1 \subset\subset \omega$ be a nonempty ball with the center x_0 and the radius τ_1. Denote by $\text{dist}((s, x), (0, x_0))$ the distance between (s, x) and $(0, x_0)$.

For any $\tau > 0$, put

$$N(\tau) = \big\{(s, x) \in Q \ : \ \text{dist}((s, x), (0, x_0)) < \tau\big\}.$$

Let $0 < \tau_1 < \tau_2 < \tau_3$ satisfy that $N(\tau_3) \subset Q$ and $N(\tau_3) \cap (\{0\} \times \Omega) \subset (\{0\} \times \omega)$. Further, let ζ be a C^2-function satisfying

$$\begin{cases} 3 < \zeta < 4 & \text{in } N(\tau_1), \\ 0 < \zeta < 1 & \text{in } N(\tau_3) \backslash N(\tau_2), \\ |\nabla \zeta| > 0 & \text{in } N(\tau_3). \end{cases} \tag{2.76}$$

The construction of ζ is very easy. For example, we can choose $g : \mathbb{R} \to \mathbb{R}$ such that

$$\begin{cases} g_t(t) < 0 \text{ for all } t \in \mathbb{R}, \\ 3 < g(t) < 4 \text{ for } 0 < t < \tau_1^2, \\ 0 < g(t) < 1 \text{ when } \tau_2^2 < t < \tau_3^2. \end{cases}$$

Then $\zeta(x, s) = g(\text{dist}((s, x), (0, x_0))^2)$ is the desired function.

For parameters $\lambda, \mu > 1$, let us define the weight functions θ, ℓ and ϕ as follows:

$$\theta = e^\ell, \quad \ell = \lambda \phi, \quad \phi = e^{\mu \zeta}. \tag{2.77}$$

Next, choose a cut-off function $\chi \in C_0^\infty(N(\tau_3))$ such that

$$\begin{cases} \chi = 1 & \text{in } N(\tau_2), \\ 0 \leq \chi \leq 1 & \text{in } N(\tau_3). \end{cases} \tag{2.78}$$

Put $\hat{u} = \chi u$, where u is the solution of the Eq. (2.52). Then, $\hat{u}$ solves the following equation:

$$
\begin{cases}
\hat{u}_{ss} + \displaystyle\sum_{j,k=1}^{n} \left(h^{jk} \hat{u}_{x_j} \right)_{x_k} = \chi_{ss} u + 2\chi_s u_s \\[2mm]
\quad + \displaystyle\sum_{j,k=1}^{n} \left(h^{jk} \chi_{x_j x_k} u + h^{jk}_{x_j} \chi_{x_k} u + 2 h^{jk} \chi_{x_j} u_{x_k} \right) \quad \text{in } N(\tau_3), \\[2mm]
|\hat{u}_s| = |\nabla \hat{u}| = \hat{u} = 0 \quad\quad\quad\quad\quad \text{on } \partial N(\tau_3) \backslash (\{0\} \times \omega).
\end{cases}
\tag{2.79}
$$

Apply Lemma 2.1 to the Eq. (2.79) with

$$
m = 1 + n, \quad x_{1+n} = s, \quad (a^{jk})_{1 \le j,k \le m} = \begin{pmatrix} (h^{jk})_{1 \le j,k \le n} & 0 \\ 0 & 1 \end{pmatrix},
$$

z being replaced by $\hat{u}$ and $v = \theta \hat{u}$. Similarly to the proof of Theorem 2.1, by (2.2) and (2.76), we obtain that

$$
\lambda \mu^2 \int_{N(\tau_3)} \phi |\nabla \zeta|^2 \left(|\nabla \hat{u}|^2 + |\hat{u}_s|^2 + \lambda^2 \mu^2 \phi^2 |\nabla \zeta|^2 |\hat{u}|^2 \right) dx ds
$$

$$
\le C \lambda \mu^2 \int_{N(\tau_3)} \phi \left(\sum_{j,k=1}^{n} h^{jk} \hat{u}_{x_j} \hat{u}_{x_k} + |\hat{u}_s|^2 + \lambda^2 \mu^2 \phi^2 |\hat{u}|^2 \right) dx ds
\tag{2.80}
$$

$$
\le C \left(\int_{N(\tau_3)} \theta^2 \left| \hat{u}_{ss} + \sum_{j,k=1}^{n} \left(h^{jk} \hat{u}_{x_j} \right)_{x_k} \right|^2 dx ds + \int_{N(\tau_3)} \sum_{k=1}^{n} V_{x_k}^k dx ds \right),
$$

where Ψ is given by (2.10) and

$$
V^k = 2 \operatorname{Re} \sum_{j=1}^{n} \left[2 h^{jk} \ell_s v_{x_j} \bar{v}_s - h^{jk} \ell_{x_j} |v_s|^2 - \Psi h^{jk} (v_{x_j} \bar{v}) + h^{jk} A \ell_{x_j} |v|^2 \right.
$$

$$
\left. + \sum_{j',k'=1}^{n} \left(2 h^{jk'} h^{j'k} - h^{jk} h^{j'k'} \right) \ell_{x_j} v_{x_{j'}} \bar{v}_{x_{k'}} \right].
\tag{2.81}
$$

Step 2. Estimation of the boundary terms. Recalling that $v = \theta \hat{u}$, and noting that $|\hat{u}_s| = |\nabla \hat{u}| = \hat{u} = 0$ on $\partial N(\tau_3) \backslash (\{0\} \times \omega)$, it is easy to see that

$$
\begin{cases}
v|_{\partial N(\tau_3) \backslash (\{0\} \times \omega)} = \hat{u}|_{\partial N(\tau_3) \backslash (\{0\} \times \omega)} = 0, \\
v_s|_{\partial N(\tau_3) \backslash (\{0\} \times \omega)} = \hat{u}_s|_{\partial N(\tau_3) \backslash (\{0\} \times \omega)} = 0, \\
\nabla v|_{\partial N(\tau_3) \backslash (\{0\} \times \omega)} = \nabla \hat{u}|_{\partial N(\tau_3) \backslash (\{0\} \times \omega)} = 0.
\end{cases}
\tag{2.82}
$$

Now we estimate the terms in "$\displaystyle\int_{N(\tau_3)}\sum_{k=1}^{n}V_{x_k}^k\,dxds$" one by one (recalling (2.81) for the definition of V^k). First of all, by (2.2), (2.13) and (2.82), we have that

$$\int_{N(\tau_3)}\operatorname{Re}\sum_{j,k=1}^{n}\left(h^{jk}\ell_s v_{x_j}\bar{v}_s\right)_{x_k}dxds = \int_{\partial N(\tau_3)}\operatorname{Re}\sum_{j,k=1}^{n}h^{jk}v^j v^k\frac{\partial v}{\partial \nu}\ell_s\bar{v}_s d\partial N(\tau_3)$$

$$\leq C\lambda\mu\int_{\{0\}\times\omega}\phi(|\nabla v|^2+|v_s|^2)dx.$$

$$(2.83)$$

Similarly, we can obtain the estimates for the other terms. For the second one, it holds that

$$\int_{N(\tau_3)}\operatorname{Re}\sum_{j,k=1}^{n}\left(h^{jk}\ell_{x_j}|v_s|^2\right)_{x_k}dxds \leq C\lambda\mu\int_{\{0\}\times\omega}\phi|v_s|^2 dx. \tag{2.84}$$

For the third one, it is clear that

$$\int_{N(\tau_3)}\operatorname{Re}\sum_{j,k=1}^{n}\left(\Psi h^{jk}v_{x_j}\bar{v}\right)_{x_k}dxds \leq C\lambda\mu\int_{\{0\}\times\omega}\phi^2\left(|\nabla v|^2+\lambda^2\mu^2|v|^2\right)dx. \tag{2.85}$$

For the fourth one, we have

$$\int_{N(\tau_3)}\operatorname{Re}\sum_{j,k=1}^{n}\left(h^{jk}A\ell_{x_j}|v|^2\right)_{x_k}dxds \leq C\lambda^3\mu^3\int_{\{0\}\times\omega}\phi^3|v|^2 dx. \tag{2.86}$$

The fifth one satisfies that

$$\int_{N(\tau_3)}\operatorname{Re}\sum_{j,k,j',k'=1}^{n}\left[(2h^{jk'}h^{j'k}-h^{jk}h^{j'k'})\ell_{x_j}v_{x_{j'}}\bar{v}_{x_{k'}}\right]_{x_k}dxds$$

$$\leq C\lambda\mu\int_{\{0\}\times\omega}\phi|\nabla v|^2 dxds.$$

$$(2.87)$$

By (2.21), (2.81) and (2.83)–(2.87), noting that $v = \theta\hat{u}$, we have that

$$\int_{N(\tau_3)}\sum_{k=1}^{n}V_{x_k}^k\,dxds \leq C\lambda\mu\int_{\{0\}\times\omega}\phi\left(|\nabla v|^2+|v_s|^2+\lambda^2\mu^2\phi^2|v|^2\right)dxds$$

$$\leq C\lambda\mu\int_{\{0\}\times\omega}\phi\left(|\nabla\hat{u}|^2+|\hat{u}_s|^2+\lambda^2\mu^2\phi^2|\hat{u}|^2\right)dxds.$$

$$(2.88)$$

Step 3. End of the proof. It follows from (2.79), (2.80) and (2.88) that

$$
\lambda\mu^2 \int_{N(\tau_3)} \phi\big(|\nabla\hat{u}|^2 + |\hat{u}_s|^2 + \lambda^2\mu^2\phi^2|\hat{u}|^2\big)dxds
$$
$$
\leq C\Big[\int_{N(\tau_3)} \theta^2\Big|\chi_{ss}u + 2\chi_s u_s + \sum_{j,k=1}^n \big(h^{jk}\chi_{x_j x_k}u + h^{jk}_{x_k}\chi_{x_j}u + 2h^{jk}\chi_{x_j}u_{x_k}\big)\Big|^2 dxds
$$
$$
+ \lambda\mu \int_{\{0\}\times\omega} \phi\big(|\nabla\hat{u}|^2 + |\hat{u}_s|^2 + \lambda^2\mu^2\phi^2|\hat{u}|^2\big)dx\Big]. \tag{2.89}
$$

It is easy to see from (2.76) and (2.77) that

$$
\begin{cases}
\phi > e^{3\mu}, & \theta > e^{\lambda e^{3\mu}} \quad \text{in } N(\tau_1), \\
\phi < e^{\mu}, & \theta < e^{\lambda e^{\mu}} \quad \text{in } N(\tau_3)\backslash N(\tau_2), \\
\phi < e^{4\mu}, & \theta < e^{\lambda e^{4\mu}} \quad \text{in } \{0\}\times\omega.
\end{cases} \tag{2.90}
$$

On the other hand, it follows from (2.78) that $\chi_s = 0$ and $\chi_{x_j} = 0$ $(j = 1,\ldots,n)$ when $(x,s) \in N(\tau_3)\backslash N(\tau_2)$. Noting that $\hat{u} = \chi u$ and $N(\tau_1) \subset N(\tau_2) \subset N(\tau_3)$, it is easy to see that $\hat{u} = u$ in $N(\tau_1)$. Therefore, by (2.89) and (2.90), we see that

$$
\lambda\mu^2 e^{3\mu}e^{2\lambda e^{3\mu}} \int_{N(\tau_1)} \big(|\nabla u|^2 + |u_s|^2 + \lambda^2\mu^2 e^{6\mu}|u|^2\big)dxds
$$
$$
\leq \lambda\mu^2 \int_{N(\tau_3)} \phi\Big(\sum_{j,k=1}^n h^{jk}v_{x_j}v_{x_k} + |v_s|^2 + \lambda^2\mu^2\phi^2|v|^2\Big)dxds \tag{2.91}
$$
$$
\leq C\Big[e^{2\lambda e^{\mu}} \int_{N(\tau_3)} \big(|u|^2 + |\nabla u|^2 + |u_s|^2\big)dxds
$$
$$
+ \lambda\mu e^{4\mu}e^{2\lambda e^{4\mu}} \int_{\{0\}\times\omega} \big(|\nabla u|^2 + |u_s|^2 + \lambda^2\mu^2 e^{8\mu}|u|^2\big)dx\Big].
$$

If $u(0,x) = 0$ in Ω, we obtain from (2.91) that

$$
\lambda\mu^2 e^{3\mu}e^{2\lambda e^{3\mu}} \int_{N(\tau_1)} \big(|\nabla u|^2 + |u_s|^2 + \lambda^2\mu^2 e^{6\mu}|u|^2\big)dxds
$$
$$
\leq C\Big[e^{2\lambda e^{\mu}} \int_{N(\tau_3)} \big(|u|^2 + |\nabla u|^2 + |u_s|^2\big)dxds + \lambda\mu e^{12\mu}e^{2\lambda e^{4\mu}} \int_\omega |u_s(0)|^2 dx\Big].
$$

Hence, there exist $\beta > 0$ and $\varepsilon_0 > 0$ such that for any $\varepsilon \in (0, \varepsilon_0]$, it holds that

$$
|u|^2_{H^1(N(\tau_1))} \leq \varepsilon^{-\beta}|u_s(0)|^2_{L^2(\omega)} + C\varepsilon|u|^2_{H^1((0,4)\times\Omega)}, \tag{2.92}
$$

which in turn implies that the inequality (2.92) holds for every $\varepsilon > 0$.

Since $\tau_1 > 0$, there is an open ball $\mathscr{B} \subset N(\tau_1)$. It follows from (2.92) that

$$
|u|^2_{H^1(\mathscr{B})} \leq \varepsilon^{-\beta}|u_s(0)|^2_{L^2(\omega)} + C\varepsilon|u|^2_{H^1((0,4)\times\Omega)}. \tag{2.93}
$$

Let $\delta' = \dfrac{1}{1+\beta}$ and $\varepsilon = \left(\dfrac{|u_s(0)|_{L^2(\omega)}}{|u|_{H^1((0,4)\times\Omega)}}\right)^{2\delta'}$ in (2.93). Then one concludes that

$$|u|_{H^1(\mathscr{B})} \le C |u_s(0)|^{\delta'}_{L^2(\omega)} |u|^{1-\delta'}_{H^1((0,4)\times\Omega)}. \tag{2.94}$$

It suffices to show that, for any given $\mathscr{K} \subset\subset (0,4) \times \Omega$, there exist two constants $0 < \delta'' < 1$ and $C > 0$ so that

$$|u|_{H^1(\mathscr{K})} \le C |u|^{\delta''}_{H^1(\mathscr{B})} |u|^{1-\delta''}_{H^1((0,4)\times\Omega)}. \tag{2.95}$$

Indeed, by (2.94) and (2.95), we deduce that for any given subset $\mathscr{K} \subset\subset (0,4) \times \Omega$,

$$|u|_{H^1(\mathscr{K})} \le C |u_s(0)|^{\delta}_{L^2(\omega)} |u|^{1-\delta}_{H^1((0,4)\times\Omega)}, \tag{2.96}$$

where $\delta = \delta'\delta''$. Then, Lemma 2.5 follows by choosing $\mathscr{K} = (\gamma, 4-\gamma) \times \omega$.

Now let us prove (2.95). Let $\mathscr{B}_1$, $\mathscr{B}_2$ and $\mathscr{B}_3$ be three open balls in $(0,4) \times \Omega$ such that $\mathscr{B}_1 \subset\subset \mathscr{B}_2 \subset\subset \mathscr{B}_3 \subset\subset (0,4) \times \Omega$. Choose a cut-off function $\eta \in C_0^\infty((0,4) \times \Omega; [0,1])$ such that $\eta = 1$ in $\mathscr{B}_3$. Let $y = \eta u$. Then, y solves

$$\begin{cases} y_{tt} + \displaystyle\sum_{j,k=1}^{n} \left(h^{jk} y_{x_j}\right)_{x_k} = \eta_{ss} u + 2\eta_s u_s \\ \quad + \displaystyle\sum_{j,k=1}^{n} h^{jk}\left(\eta_{x_j x_k} u + 2\eta_{x_j} u_{x_k}\right) + \displaystyle\sum_{j,k=1}^{n} h^{jk}_{x_j} \eta u_{x_k} \quad \text{in } (0,4) \times \Omega, \\ \dfrac{\partial y}{\partial v} = y = 0 \qquad\qquad\qquad\qquad \text{on} \quad [(0,4) \times \Gamma] \cup [\{0,4\} \times \Omega]. \end{cases} \tag{2.97}$$

Denote by $\mathbf{p}$ the center of $\mathscr{B}_1$. Let $\tilde\varphi(t,x) = \text{dist}((t,x), \mathbf{p})^2$. Replace φ in θ by $\tilde\varphi$. Proceeding as in the proof of Lemma 2.3, one can find two constants $C > 0$ and $0 < \tilde\delta < 1$ such that

$$|u|_{H^1(\mathscr{B}_2)} \le C |u|^{\tilde\delta}_{H^1(\mathscr{B}_1)} |u|^{1-\tilde\delta}_{H^1((0,4)\times\Omega)}. \tag{2.98}$$

For any ball $\mathscr{B}' \subset\subset (0,4) \times \Omega$, there exist a finite number $l \in \mathbb{N}$ and two sequences of balls $\{\mathscr{B}^j\}_{j=1}^{l}$ and $\{\tilde{\mathscr{B}}^j\}_{j=1}^{l}$ such that

$$\begin{cases} \mathscr{B}' \subset\subset \mathscr{B}^1, \\ \tilde{\mathscr{B}}^j \subset\subset \mathscr{B}^j \cap \mathscr{B}^{j+1} \text{ for } j = 1,\ldots, l-1, \\ \tilde{\mathscr{B}}^l \subset\subset \mathscr{B}^l, \\ \tilde{\mathscr{B}}^l = \mathscr{B}. \end{cases}$$

Hence, from (2.98), we can find a sequences $\{\tilde\delta_j\}_{j=1}^{l}$ satisfying $0 < \tilde\delta_j < 1$ for $j = 1,\ldots, l$, such that

$$|u|_{H^1(\mathscr{B}')} \leq |u|_{H^1(\mathscr{B}^1)} \leq C|u|_{H^1(\tilde{\mathscr{B}}^1)}^{\tilde{\delta}_1}|u|_{H^1((0,4)\times\Omega)}^{1-\tilde{\delta}_1} \leq C|u|_{H^1(\mathscr{B}^2)}^{\tilde{\delta}_1}|u|_{H^1((0,4)\times\Omega)}^{1-\tilde{\delta}_1}$$
$$\leq C|u|_{H^1(\tilde{\mathscr{B}}^2)}^{\tilde{\delta}_1\tilde{\delta}_2}|u|_{H^1((0,4)\times\Omega)}^{1-\tilde{\delta}_1\tilde{\delta}_2} \leq \cdots \leq C|u|_{H^1(\tilde{\mathscr{B}}^\ell)}^{\tilde{\delta}_1\tilde{\delta}_2\cdots\tilde{\delta}_l}|u|_{H^1((0,4)\times\Omega)}^{1-\tilde{\delta}_1\tilde{\delta}_2\cdots\tilde{\delta}_l}.$$

Put $\tilde{\tilde{\delta}} = \tilde{\delta}_1\tilde{\delta}_2\cdots\tilde{\delta}_l$. Then

$$|u|_{H^1(\mathscr{B}')} \leq C|u|_{H^1(\mathscr{B})}^{\tilde{\tilde{\delta}}}|u|_{H^1((0,4)\times\Omega)}^{1-\tilde{\tilde{\delta}}}. \tag{2.99}$$

For any given $\mathscr{K} \subset\subset (0,4) \times \Omega$, we may find finite many balls contained in $(0,4) \times \Omega$ to cover it. By (2.99), we deduce that (2.95) holds for suitable constants $\delta'' \in (0,1)$ and $C > 0$. This completes the proof of Lemma 2.5.

2.3   Logarithmic Decay of Locally Damped Hyperbolic Equations

As another application of Theorem 2.1, in this section, we shall analyze the longtime behavior of hyperbolic equations with damping acted on an arbitrary small nonempty open subset of Ω.

Let $a(\cdot) \in L^\infty(\Omega; \mathbb{R})$ be a non-negative function satisfying

$$a(x) \geq a_0 > 0 \qquad \text{for a.e. } x \in \omega, \tag{2.100}$$

where a_0 is a given constant.

Consider the following damped hyperbolic equation:

$$\begin{cases} u_{tt} - \displaystyle\sum_{j,k=1}^n (h^{jk}u_{x_j})_{x_k} + a(x)u_t = 0 & \text{in } \mathbb{R}^+ \times \Omega, \\ u = 0 & \text{on } \mathbb{R}^+ \times \Gamma, \\ u(0) = u^0, \ u_t(0) = u^1 & \text{in } \Omega. \end{cases} \tag{2.101}$$

Put $\mathbb{H} \overset{\triangle}{=} H_0^1(\Omega) \times L^2(\Omega)$, which is a Hilbert space with the norm given by

$$|(f,g)|_{\mathbb{H}} = \sqrt{\int_\Omega \left(\sum_{j,k=1}^n h^{jk} f_{x_j}\overline{f_{x_k}} + |g|^2 \right) dx}, \qquad \forall\, (f,g) \in \mathbb{H}.$$

Define an unbounded operator $\mathscr{A} : \mathbb{H} \to \mathbb{H}$ by

$$\begin{cases} \mathscr{A} \overset{\triangle}{=} \begin{pmatrix} 0 & I \\ \displaystyle\sum_{j,k=1}^{n} \partial_{x_k}(h^{jk}\partial_{x_j}) & -a(x)I \end{pmatrix}, \\ D(\mathscr{A}) \overset{\triangle}{=} (H^2(\Omega) \cap H_0^1(\Omega)) \times H_0^1(\Omega). \end{cases}$$

It is easy to show that $\mathscr{A}$ generates a C_0-semigroup $\{e^{t\mathscr{A}}\}_{t\in\mathbb{R}}$ on $\mathbb{H}$. Multiplying the first equation in (2.101) by $\bar{u}_t$ and integrating the resulting equality in Ω, using integration by parts, we find that

$$\frac{d}{dt}|(u, u_t)|_{\mathbb{H}}^2 = -2\int_{\Omega} a(x)|u_t|^2 dx.$$

This indicates that $|(u, u_t)|_{\mathbb{H}}^2$ decays as t increase.

The following result provides the decay rate of solutions to the system (2.101).

Theorem 2.4 *There is a constant $C > 0$ such that for any $(u^0, u^1) \in D(\mathscr{A})$, the corresponding solution $(u, u_t) \in C([0, \infty); D(\mathscr{A})) \cap C^1([0, \infty); \mathbb{H})$ to (2.101) satisfies that*

$$\left|(u(t), u_t(t))\right|_{\mathbb{H}} \le \frac{C}{\ln(2+t)}|(u^0, u^1)|_{D(\mathscr{A})}, \quad \forall t > 0.$$

Remark 2.13 Logarithmic decay results for damped hyperbolic equations in bounded domains with C^∞ boundaries were first considered in [31, 33], in which the authors employed a local Carleman estimate. Theorem 2.4 was proved in [13]. A similar result for hyperbolic equations with damping acting on the boundary was obtained in [12]. The generalization to systems of hyperbolic equations can be found in [14]. Here, we only consider the logarithmic decay problem, while the exponential decay problem will be discussed in Chap. 4.

Denote by I the identity operator on $\mathbb{H}$, and by $\mathscr{L}(\mathbb{H})$ the Banach space of all bounded linear operators from $\mathbb{H}$ to itself, with the usual operator norm. It is well known that, once a suitable resolvent estimate for the operator $\mathscr{A}$ is established, the existing results for C_0-semigroup can be adopted to yield the desired energy decay rate (see [5, Théorème 3] or [38, Theorem 2.1]). Hence, to prove Theorem 2.4, we only need to establish the following resolvent estimate:

Theorem 2.5 *There exists a constant $C > 0$ such that for any $\lambda \in \mathbb{C}$ satisfying*

$$\operatorname{Re} \lambda \in \left[-e^{-C|\operatorname{Im}\lambda|}/C, 0 \right],$$

it holds that

$$|(\mathscr{A} - \lambda I)^{-1}|_{\mathscr{L}(\mathbb{H})} \le Ce^{C|\operatorname{Im}\lambda|}, \quad \text{for } |\lambda| > 1.$$

Proof We divide the proof into two steps.

Step 1. First, fix $f = (f^0, f^1) \in \mathbb{H}$ and $u = (u^0, u^1) \in D(\mathscr{A})$. It is easy to see that

$$(\mathscr{A} - \lambda I)u = f$$

is equivalent to

$$\begin{cases} -\lambda u^0 + u^1 = f^0 & \text{in } \Omega, \\ \displaystyle\sum_{j,k=1}^{n} (h^{jk} u^0_{x_j})_{x_k} - (a + \lambda)u^1 = f^1 & \text{in } \Omega. \end{cases} \tag{2.102}$$

By (2.102) and noting the boundary condition, we find

$$\begin{cases} \displaystyle\sum_{j,k=1}^{n} (h^{jk} u^0_{x_j})_{x_k} - \lambda^2 u^0 - \lambda a u^0 = (a + \lambda) f^0 + f^1 & \text{in } \Omega, \\ u^0 = 0 & \text{on } \Gamma, \\ u^1 = f^0 + \lambda u^0 & \text{in } \Omega. \end{cases}$$

Put $v = e^{i\lambda s} u^0$. Then v solves

$$\begin{cases} \displaystyle v_{ss} + \sum_{j,k=1}^{n} (h^{jk} v_{x_j})_{x_k} + i a v_s = \left(\lambda f^0 + a f^0 + f^1\right) e^{i\lambda s} & \text{in } \mathbb{R} \times \Omega, \\ v = 0 & \text{on } \mathbb{R} \times \Gamma. \end{cases} \tag{2.103}$$

Put

$$v^0 \overset{\triangle}{=} (\lambda f^0 + a f^0 + f^1) e^{i\lambda s}. \tag{2.104}$$

Choose $g = v^0 - ia(x)v_s$ in (2.56). By (2.55), for all $\varepsilon > 0$ and every solution $v \in H^2((-2,2) \times \Omega; \mathbb{C})$ of (2.103), it holds that

$$\begin{aligned} |v|_{H^1((-1,1)\times\Omega)} &\leq C e^{C/\varepsilon} \big(|v^0 - ia(x)v_s|_{L^2((-2,2)\times\Omega)} + |v|_{L^2((-2,2)\times\omega)} \big) \\ &\quad + C e^{-2/\varepsilon} |v|_{H^1((-2,2)\times\Omega)}. \end{aligned} \tag{2.105}$$

Combining (2.105) with (2.100), we conclude that

$$\begin{aligned} |v|_{H^1((-1,1)\times\Omega)} &\leq C e^{C/\varepsilon} \big(|v^0|_{L^2((-2,2)\times\Omega)} + |v|_{L^2((-2,2)\times\omega)} + |v_s|_{L^2((-2,2)\times\omega)} \big) \\ &\quad + C e^{-2/\varepsilon} |v|_{H^1((-2,2)\times\Omega)}. \end{aligned} \tag{2.106}$$

Step 2. Clearly,

$$\begin{cases} |u^0|_{H^1(\Omega)} \leq C e^{C|\operatorname{Im}\lambda|} |v|_{H^1((-1,1)\times\Omega)}, \\ |v|_{H^1((-2,2)\times\Omega)} \leq C(|\lambda| + 1) e^{C|\operatorname{Im}\lambda|} |u^0|_{H^1(\Omega)}, \\ |v|_{L^2((-2,2)\times\omega)} \leq C e^{C|\operatorname{Im}\lambda|} |u^0|_{L^2(\omega)}, \\ |v_s|_{L^2((-2,2)\times\omega)} \leq C|\lambda| e^{C|\operatorname{Im}\lambda|} |u^0|_{L^2(\omega)}. \end{cases} \tag{2.107}$$

From (2.104), (2.106) and (2.107), we get that

$$|u^0|_{H^1(\Omega)} \leq Ce^{C|\operatorname{Im}\lambda|}\big(|f^0|_{H^1(\Omega)} + |f^1|_{L^2(\Omega)} + |u^0|_{L^2(\omega)}\big). \tag{2.108}$$

On the other hand, multiplying (2.102) by $\overline{u}^0$ and integrating it on Ω, we obtain that

$$\int_{\Omega}\Big[-\sum_{j,k=1}^{n}(h^{jk}u^0_{x_j})_{x_k} + \lambda^2 u^0 + \lambda a u^0\Big]\overline{u}^0 dx$$
$$= \lambda^2|u^0|^2_{L^2(\Omega)} + \sum_{j,k=1}^{n}\int_{\Omega} h^{jk}u^0_{x_j}\overline{u}^0_{x_k}dx + \int_{\Omega} a\lambda|u^0|^2 dx. \tag{2.109}$$

By taking the imaginary part in the both sides of (2.109), we find that

$$|\operatorname{Im}\lambda|\int_{\Omega} a|u^0|^2 dx$$
$$\leq \Big| -\sum_{j,k=1}^{n}(h^{jk}u^0_{x_j})_{x_k} + \lambda^2 u^0 + \lambda a u^0\Big|_{L^2(\Omega)}|u^0|_{L^2(\Omega)} + 2|\operatorname{Im}\lambda||\operatorname{Re}\lambda||u^0|^2_{L^2(\Omega)}$$
$$\leq C\Big(|\lambda f^0 + a f^0 + f^1|_{L^2(\Omega)}|u^0|_{L^2(\Omega)} + |\operatorname{Im}\lambda||\operatorname{Re}\lambda||u^0|^2_{L^2(\Omega)}\Big). \tag{2.110}$$

Combining (2.108) and (2.110), we have that

$$|u^0|_{H^1(\Omega)} \leq Ce^{C|\operatorname{Im}\lambda|}\Big(|f^0|_{H^1(\Omega)} + |f^1|_{L^2(\Omega)} + |\operatorname{Re}\lambda||u^0|_{H^1(\Omega)}\Big). \tag{2.111}$$

Take

$$Ce^{C|\operatorname{Im}\lambda|}|\operatorname{Re}\lambda| \leq \frac{1}{2},$$

which holds, whenever $|\operatorname{Re}\lambda| \leq \frac{1}{C}e^{-C|\operatorname{Im}\lambda|}$ for some sufficiently large $C > 0$. Then, by (2.111), we have that

$$|u^0|_{H^1(\Omega)} \leq Ce^{C|\operatorname{Im}\lambda|}\big(|f^0|_{H^1(\Omega)} + |f^1|_{L^2(\Omega)}\big). \tag{2.112}$$

By $u^1 = f^0 + \lambda u^0$, it follows

$$|u^1|_{L^2(\Omega)} \leq |f^0|_{L^2(\Omega)} + |\lambda||u^0|_{L^2(\Omega)} \leq Ce^{C|\operatorname{Im}\lambda|}\big(|f^0|_{H^1(\Omega)} + |f^1|_{L^2(\Omega)}\big). \tag{2.113}$$

Combining (2.112) and (2.113), we obtain

$$|(\mathscr{A} - \lambda I)^{-1}|_{\mathscr{L}(\mathbb{H})} \leq Ce^{C|\operatorname{Im}\lambda|}, \quad \text{for } \operatorname{Re}\lambda \in [-e^{C|\operatorname{Im}\lambda|}/C, 0], \quad |\lambda| \geq 1.$$

This completes the proof of Theorem 2.5.

2.4 Strong Unique Continuation Property of Elliptic Equations

As an application of Theorem 2.2, in this section we study the SUCP of elliptic equations, which originates from the classical work of Carleman [6].

Consider the following equation:

$$-\sum_{j,k=1}^{n} (h^{jk} y_{x_j})_{x_k} + by = 0 \qquad \text{in } \Omega, \tag{2.114}$$

where $b \in L^{\infty}_{loc}(\Omega; \mathbb{C})$.

Definition 2.1 A solution y to (2.114) is said to satisfy the SUCP if $y = 0$ in Ω whenever it vanishes of infinite order at some $x_0 \in \Omega$.

Theorem 2.6 *Let $y \in H^2_{loc}(\Omega; \mathbb{C})$ be a solution of (2.114). Then y satisfies the SUCP.*

Remark 2.14 A natural generalization of the SUCP might be as follows: $y = 0$ in Ω provided that, there exist $x_0 \in \Omega$, $N_0 \in \mathbb{N}$, $C > 0$ and $R > 0$ such that $\int_{\mathcal{B}(x_0,r)} |y|^2 dx \le Cr^{2N}$, for all $N \le N_0$ and $r \in (0, R)$. Unfortunately, such kind of result cannot be true. A counterexample is $y(x_1, x_2) = \text{Re}\,(x_1 + ix_2)^{N_0+1}$ for any given $N_0 \in \mathbb{N}$. Nevertheless, one can obtain some positive results under some more restrictions on the solutions. A detailed introduction to this topic is beyond the scope of this book. Interested readers are referred to [4, 9] for this topic.

Remark 2.15 In (2.114), we assume that $b \in L^{\infty}_{loc}(\Omega; \mathbb{C})$. A sharp condition for SUCP is that $b \in L^{n/2}_{loc}(\Omega; \mathbb{C})$ (See [23] for the positive result and [26] for the counterexample). One need to employ an L^p-Carleman estimate (for a suitable $p \ge 1$) to establish such a result. In [27], when $b \in L^{\frac{n+1}{2}}_{loc}(\Omega; \mathbb{C})$, an L^2-Carleman estimate (as that in this chapter) and a dispersive estimate were combined to show SUCP for (2.114). Until now, we do not know how to handle the case $b \in L^{n/2}_{loc}(\Omega; \mathbb{C})$ by the techniques in this book.

Remark 2.16 As mentioned in Remark 2.1, by modifying the proof in this section, one can show that SUCP still holds if h^{jk} is Lipschitz continuous for $j, k = 1, \ldots, n$. This result is sharp. Indeed, [46] showed the existence of a nonzero solution y which vanishes in an open set, whenever a^{jk} ($j, k = 1, \ldots, n$) is Hölder continuous of any order less than 1.

Before proving Theorem 2.6, we first recall the following Caccioppoli inequality.

Lemma 2.6 *Let $0 < s_1 < s_2$. Then, there exists a constant $C > 0$, depending only on Ω, s_1, s_2 and h_0, such that*

$$h_0 \int_{\mathscr{B}(x_0, s_1 r)} |\nabla y|^2 dx$$

$$\leq \frac{C}{(s_2 - s_1)^2 r^2} \int_{\mathscr{B}(x_0, s_2 r)} |y|^2 dx + |b|_{L^\infty(\mathscr{B}(x_0, s_2 r))} \int_{\mathscr{B}(x_0, s_2 r)} |y|^2 dx, \tag{2.115}$$

for any $x_0 \in \Omega$, $0 < r < \frac{1}{s_2} \text{dist}(x_0, \Gamma)$ and $y \in H^1_{loc}(\Omega)$ solving (2.114).

Proof Let $\rho \in C^\infty_0(\Omega; [0, 1])$ be such that

$$\begin{cases} \rho \equiv 1 & \text{in } \mathscr{B}(x_0, s_1 r), \\ \rho \equiv 0 & \text{in } \Omega \setminus \mathscr{B}(x_0, s_2 r), \\ |\nabla \rho| \leq \dfrac{2}{s_2 r - s_1 r}. \end{cases}$$

Multiplying (2.114) by $\rho^2 \bar{y}$ and using integration by parts, we have

$$\int_{\mathscr{B}(x_0, s_2 r)} \rho^2 h^{jk} y_{x_j} \bar{y}_{x_k} dx$$

$$= - \int_{\mathscr{B}(x_0, s_2 r)} \rho^2 b |y|^2 dx + \int_{\mathscr{B}(x_0, s_2 r)} \sum_{j,k=1}^n h^{jk} y_{x_j} (2\rho \rho_{x_k} \bar{y}) dx$$

$$\leq |b|_{L^\infty(\mathscr{B}(x_0, s_2 r))} \int_{\mathscr{B}(x_0, s_2 r)} |y|^2 dx$$

$$+ \left(\int_{\mathscr{B}(x_0, s_2 r)} \sum_{j,k=1}^n h^{jk} y_{x_j} \bar{y}_{x_k} \rho^2 dx \right)^{\frac{1}{2}} \left(\int_{\mathscr{B}(x_0, s_2 r)} 4 |y|^2 \sum_{j,k=1}^n h^{jk} \rho_{x_j} \rho_{x_k} dx \right)^{\frac{1}{2}}. \tag{2.116}$$

Then, by (2.2) and (2.116), we obtain the desired result immediately.

In the sequel, for any $x_0 \in \mathbb{R}^n$ and $0 < \delta_1 < \delta_2$, set

$$\mathscr{D}(x_0, \delta_1, \delta_2) \stackrel{\triangle}{=} \mathscr{B}(x_0, \delta_2) \setminus \mathscr{B}(x_0, \delta_1).$$

We are now in a position to prove Theorem 2.6.

Proof of Theorem 2.6. Without loss of generality, we assume that $x_0 = 0$. Let $\delta_0 > 0$ be such that $\mathscr{B}(0, \delta_0) \subset \Omega$. Let us first introduce two cut-off functions:

$$\rho_1 \in C^\infty_0(\mathscr{B}(0, 3\delta_0/4), [0, 1]), \quad \rho_1 \equiv 1 \text{ on } B(0, \delta_0/2)$$

and

$$\rho_2 \in C^\infty(\mathbb{R}^n), \quad \rho_2 \equiv 1 \text{ in } \mathbb{R}^n \setminus \mathscr{B}(0, 1), \quad \rho_2 \equiv 0 \text{ in } \mathscr{B}(0, 1/2).$$

For $\varepsilon \in \left(0, \min\{1, \delta_0/2\} \right)$, put

$$\eta_\varepsilon(x) = \rho_2 \left(\frac{x}{\varepsilon} \right) \rho_1(x), \quad y_\varepsilon(x) = \eta_\varepsilon(x) y(x). \tag{2.117}$$

By Theorems 2.2 and (2.114), we obtain that

$$
\lambda^3 \int_{\mathscr{B}(0,\delta_0)} w^{-1-2\lambda} |y_\varepsilon|^2 dx
$$

$$
\leq C \int_{\mathscr{B}(0,\delta_0)} w^{2-2\lambda} \left| \sum_{j,k=1}^{n} (h^{jk} y_{\varepsilon,x_j})_{x_k} \right|^2 dx
\tag{2.118}
$$

$$
\leq C \int_{\mathscr{B}(0,\delta_0)} w^{2-2\lambda} \left| \eta_\varepsilon by + \sum_{j,k=1}^{n} \left[2h^{jk} y_{x_j} \eta_{\varepsilon,x_k} + y\left(h^{jk}\eta_{\varepsilon,x_j}\right)_{x_k} \right] \right|^2 dx,
$$

where $w(\cdot)$ is given by (2.22).

It follows from (2.117) that

$$
\eta_{\varepsilon,x_k}(x) = \frac{1}{\varepsilon} \rho_{2,x_k}\left(\frac{x}{\varepsilon}\right) \rho_1(x) + \rho_2\left(\frac{x}{\varepsilon}\right)\rho_{1,x_k}(x),
$$

and that

$$
\eta_{\varepsilon,x_j x_k}(x) = \frac{1}{\varepsilon^2} \rho_{2,x_j x_k}\left(\frac{x}{\varepsilon}\right)\rho_1(x) + \rho_2\left(\frac{x}{\varepsilon}\right)\rho_{1,x_j x_k}(x)
$$
$$
+ \frac{1}{\varepsilon}\rho_{2,x_j}\left(\frac{x}{\varepsilon}\right)\rho_{1,x_k}(x) + \frac{1}{\varepsilon}\rho_{2,x_k}\left(\frac{x}{\varepsilon}\right)\rho_{1,x_j}(x).
$$

Hence,

$$
\int_{\mathscr{B}(0,\delta_0)} w^{2-2\lambda} \left| \sum_{j,k=1}^{n} h^{jk} y_{x_j} \eta_{\varepsilon,x_k} \right|^2 dx
$$

$$
\leq 2 \int_{\mathscr{B}(0,\delta_0)} w^{2-2\lambda} \left| \frac{1}{\varepsilon} \sum_{j,k=1}^{n} h^{jk} y_{x_j} \rho_{2,x_k}\left(\frac{x}{\varepsilon}\right)\rho_1(x) \right|^2 dx
$$

$$
+ 2 \int_{\mathscr{B}(0,\delta_0)} w^{2-2\lambda} \left| \sum_{j,k=1}^{n} h^{jk} y_{x_j} \rho_2\left(\frac{x}{\varepsilon}\right)\rho_{1,x_k}(x) \right|^2 dx.
$$

By (2.46) and (2.115), we obtain that

$$
\int_{\mathscr{B}(0,\delta_0)} w^{2-2\lambda} \left| \frac{1}{\varepsilon} \sum_{j,k=1}^{n} h^{jk} y_{x_j} \rho_{2,x_k}\left(\frac{x}{\varepsilon}\right)\rho_1(x) \right|^2 dx
$$

$$
\leq C \int_{\mathscr{D}(0,\varepsilon/2,\varepsilon)} \frac{1}{\varepsilon^2} w^{2-2\lambda} |\nabla y|^2 dx
$$

$$
\leq \frac{C}{\varepsilon^2} \left(\frac{\varepsilon}{2C_1}\right)^{2-2\lambda} \int_{\mathscr{D}(0,\varepsilon/2,\varepsilon)} |\nabla y|^2 dx
\tag{2.119}
$$

$$
\leq \frac{C}{\varepsilon^2} \left(\frac{\varepsilon}{2C_1}\right)^{2-2\lambda} \left(|b|_{L^\infty(\mathscr{B}(0,2\varepsilon))} + \frac{C}{\varepsilon^2} \right) \int_{\mathscr{D}(0,\varepsilon/4,2\varepsilon)} |y|^2 dx
$$

$$
\leq C C_N \left(|b|_{L^\infty(\mathscr{B}(0,\delta_0))} + 1 \right) C_1^{2\lambda-2} 2^{2\lambda-2+2N} \varepsilon^{2N-2\lambda-2},
$$

where we have used the fact that $\int_{\mathscr{B}(0,r)} |y|^2 dx \leq C_N r^{2N}$ for all $N \geq 0$.

Let us choose N to be large enough such that the right hand side of (2.119) tends to 0 as $\varepsilon \to 0$. Furthermore, proceeding exactly the same analysis as that in (2.119), it is easy to see that all the terms in which ρ_2 is differentiated on the right hand side of (2.118) tends to 0 as $\varepsilon \to 0$. Consequently, letting $\varepsilon \to 0$ in (2.118), we obtain that

$$
\begin{aligned}
&\lambda^3 \int_{\mathscr{B}(0,\delta_0)} w^{-1-2\lambda} |\rho_1 y|^2 dx \\
&\leq C \int_{\mathscr{B}(0,\delta_0)} w^{2-2\lambda} \left| \rho_1 by + y \sum_{j,k=1}^{n} (h^{jk} \rho_{1,x_j})_{x_k} + 2 \sum_{j,k=1}^{n} h^{jk} y_{x_j} \rho_{1,x_k} \right|^2 dx.
\end{aligned}
\tag{2.120}
$$

By (2.46), we get that

$$
w(x)^{2-2\lambda} = w(x)^3 w(x)^{-1-2\lambda} \leq (C_1\delta_0)^3 w(x)^{-1-2\lambda}, \qquad \forall x \in \mathscr{B}(0,\delta_0).
$$

Let λ be large enough such that

$$
C\delta_0^3 |b|^2_{L^\infty(\mathscr{B}(0,\delta_0))} \leq \frac{\lambda^3}{2}.
$$

It follows from (2.46), (2.120), (2.23) and the monotonicity of $\tilde{\varphi}$ in (2.22) that

$$
\begin{aligned}
&\frac{\lambda^3}{2} \int_{\mathscr{B}(0,\delta_0)} w^{-1-2\lambda} |\rho_1 y|^2 dx \\
&\leq C \int_{\mathscr{B}(0,\delta_0)} w^{2-2\lambda} \left| y \sum_{j,k=1}^{n} (h^{jk} \rho_{1,x_j})_{x_k} + 2 \sum_{j,k=1}^{n} h^{jk} y_{x_j} \rho_{1,x_k} \right|^2 dx \\
&\leq C \tilde{\varphi}^{2-2\lambda}\left(\frac{\delta_0}{2}\right) \int_{\mathscr{D}(0,\delta_0/2,3\delta_0/4)} (|y|^2 + |\nabla y|^2) dx \\
&\leq C(|b|_{L^\infty(\mathscr{B}(0,\delta_0))} + 1) \tilde{\varphi}^{-1-2\lambda}\left(\frac{\delta_0}{2}\right) \int_{\mathscr{B}(0,\delta_0)} |y|^2 dx \\
&\leq C(|b|_{L^\infty(\mathscr{B}(0,\delta_0))} + 1) C_0^2 \tilde{\varphi}^{-1-2\lambda}\left(\frac{\delta_0}{2}\right),
\end{aligned}
$$

where we have used Lemma 2.6 once again. Consequently,

$$
\frac{\lambda^3}{2} \tilde{\varphi}\left(\frac{\delta_0}{2}\right)^{1+2\lambda} \int_{\mathscr{B}(0,\delta_0/2)} w(x)^{-1-2\lambda} |y|^2 dx \leq CC_0^2 (|b|_{L^\infty(\mathscr{B}(0,\delta_0))} + 1).
\tag{2.121}
$$

Since $w(x) = \tilde{\varphi}(|x|)$ and $\tilde{\varphi}$ is an increasing function, $w(x)\tilde{\varphi}(\frac{\delta_0}{2})^{-1} < 1$ in $\mathscr{B}(0, \frac{\delta_0}{2})$. By letting $\lambda \to \infty$, it follows from (2.121) that $y = 0$ in $\mathscr{B}(0, \frac{\delta_0}{2})$. Then, similar to the proof of Lemma 2.5, a chain of balls argument shows that $y = 0$ in Ω.

2.5 Three-Ball Inequality of Elliptic Equations

As another application of Theorem 2.2, in this section we establish a three-ball inequality for the following elliptic equation:

$$-\sum_{j,k=1}^{n} (h^{jk}z_{x_j})_{x_k} + cz = 0 \quad \text{in } \Omega, \tag{2.122}$$

where $c \in L^\infty(\Omega; \mathbb{C})$.

Without loss of generality, we assume that $x_0 = 0 \in \Omega$. Let $\delta_0 > 0$ be such that $\mathscr{B}(0, \delta_0) \subset \Omega$. For simplicity, set

$$r^* = \frac{\delta_0}{4}.$$

We have the following three-ball inequality for solutions to the elliptic equation (2.122).

Theorem 2.7 *Let $z \in L^2_{loc}(\Omega; \mathbb{C})$ be a solution to (2.122) with $|c|_{L^\infty(\Omega)} \le M$ for some constant $M \gg 1$. Then there exists a constant $C > 0$ such that*

$$|z|_{L^2(\mathscr{B}(0,2r^*))} \le \left[CM + e^{C(\alpha+\beta)M^{2/3}} \right] |z|_{L^2(\mathscr{B}(0,r^*))}^{\frac{\alpha}{\alpha+\beta}} |z|_{L^2(\mathscr{B}(0,3r^*))}^{\frac{\beta}{\alpha+\beta}}, \tag{2.123}$$

where

$$\alpha = \ln\left[\tilde{\varphi}(2r^*)/\tilde{\varphi}(r^*/2)\right] > 0, \quad \beta = \ln\left[\tilde{\varphi}(5r^*/2)/\tilde{\varphi}(2r^*)\right] > 0 \tag{2.124}$$

and $\tilde{\varphi}(\cdot)$ is given by (2.22).

Proof Let $\zeta \in C_0^\infty(\mathscr{U})$ satisfying $0 \le \zeta \le 1$ and $\zeta = 1$ in $\mathscr{K}$, with

$$\mathscr{U} = \mathscr{D}(r^*/2, 11r^*/4), \quad \mathscr{K} = \mathscr{D}(3r^*/4, 5r^*/2). \tag{2.125}$$

By applying Theorem 2.2 (with Ω being replaced by $\mathscr{U}$) to ζz, we see that there are constants $\lambda_0 > 0$ and $C > 0$ such that for all $z \in C_0^2(\mathscr{U})$ and $\lambda \ge \lambda_0$,

$$\lambda^3 \int_{\mathscr{U}} w^{-1-2\lambda} |\zeta z|^2 dx \le C \int_{\mathscr{U}} w^{2-2\lambda} \left| \sum_{j,k=1}^{n} \left[h^{jk}(\zeta z)_{x_j} \right]_{x_k} \right|^2 dx$$

$$\le C \int_{\mathscr{U}} w^{2-2\lambda} \left| \zeta c z + z \sum_{j,k=1}^{n} (h^{jk}\zeta_{x_j})_{x_k} + 2 \sum_{j,k=1}^{n} h^{jk} z_{x_j} \zeta_{x_k} \right|^2 dx, \tag{2.126}$$

where w is the weight function given in (2.22).

Noting that $w(x) = \tilde{\varphi}(|x|)$, we have

$$w(x)^{2-2\lambda} = w(x)^3 w(x)^{-1-2\lambda} \le \tilde{\varphi}\left(\frac{11}{4}r^*\right)^3 w(x)^{-1-2\lambda}.$$

Choose λ_1 as follows

$$\tilde{\varphi}\left(\frac{11}{4}r^*\right)^4 CM^2 = \frac{\lambda_1^3}{2}. \tag{2.127}$$

Then, for any $\lambda \ge \max(\lambda_1, \lambda_0)$, it follows from (2.126) that

$$\frac{\lambda^3}{2} \int_{\mathscr{U}} w^{-1-2\lambda} |\zeta z|^2 dx \le C \int_{\mathscr{U}} w^{-1-2\lambda} \left| z \sum_{j,k=1}^n (h^{jk}\zeta_{x_j})_{x_k} + 2 \sum_{j,k=1}^n h^{jk} z_{x_j} \zeta_{x_k} \right|^2 dx.$$

Noting that $\mathscr{D}(3r^*/4, 2r^*) \subset \mathscr{U}$, by (2.126), we have that

$$\lambda^3 \int_{\mathscr{D}(3r^*/4,2r^*)} w^{-1-2\lambda} |z|^2 dx$$
$$\le C \int_{\mathscr{D}(r^*/2,3r^*/4) \cup \mathscr{D}(5r^*/2,11r^*/4)} w^{-1-2\lambda}(|\nabla z|^2 + |z|^2) dx.$$

Hence, for $\lambda \ge \lambda_1$,

$$\int_{\mathscr{D}(r^*/2,2r^*)} w^{-1-2\lambda} |z|^2 dx$$
$$\le C \int_{\mathscr{D}(r^*/2,3r^*/4)} w^{-1-2\lambda}(|\nabla z|^2 + |z|^2) dx \tag{2.128}$$
$$+ C \int_{\mathscr{D}(5r^*/2,11r^*/4)} w^{-1-2\lambda}(|\nabla z|^2 + |z|^2) dx.$$

Again, recalling (2.22) for the definition of w, by (2.128), we have that

$$\tilde{\varphi}(2r^*)^{-1-2\lambda} \int_{\mathscr{D}(r^*/2,2r^*)} |z|^2 dx$$
$$\le C\tilde{\varphi}(r^*/2)^{-1-2\lambda} \int_{\mathscr{D}(r^*/2,3r^*/4)} (|\nabla z|^2 + |z|^2) dx$$
$$+ C\tilde{\varphi}(5r^*/2)^{-1-2\lambda} \int_{\mathscr{D}(5r^*/2,11r^*/4)} (|\nabla z|^2 + |z|^2) dx, \quad \lambda \ge \lambda_1.$$

Thus, for $\lambda \ge \lambda_1$,

$$\int_{\mathcal{D}(r^*/2,2r^*)} |z|^2 dx$$

$$\leq C\tilde{\varphi}(2r^*)\tilde{\varphi}(r^*/2)\Big(\frac{\tilde{\varphi}(r^*/2)}{\tilde{\varphi}(2r^*)}\Big)^{-2\lambda} \int_{\mathcal{D}(r^*/2,3r^*/4)} \big(|\nabla z|^2 + |z|^2\big)dx \qquad (2.129)$$

$$+ C\tilde{\varphi}(2r^*)\tilde{\varphi}(5r^*/2)\Big(\frac{\tilde{\varphi}(5r^*/2)}{\tilde{\varphi}(2r^*)}\Big)^{-2\lambda} \int_{\mathcal{D}(5r^*/2,11r^*/4)} \big(|\nabla z|^2 + |z|^2\big)dx.$$

Next, by Lemma 2.6 and noting that $\delta_0 = 4r^*$, we have

$$h_0 \int_{\mathcal{D}(r^*/2,3r^*/4)} |\nabla z|^2 dx \leq \Big(M + \frac{C}{\delta_0^2}\Big) \int_{\mathcal{D}(r^*/4,r^*)} |z|^2 dx \qquad (2.130)$$

and

$$h_0 \int_{\mathcal{D}(5r^*/2,11^*/4)} |\nabla z|^2 dx \leq \Big(M + \frac{C}{\delta_0^2}\Big) \int_{\mathcal{D}(9r^*/4,3r^*)} |z|^2 dx. \qquad (2.131)$$

Now, adding $\int_{\mathcal{B}(0,r^*/2)} |z|^2 dx$ to both sides of the inequality (2.129), by (2.130) and (2.131), for any $\lambda \geq \lambda_1$, we have that

$$\int_{\mathcal{B}(0,2r^*)} |z|^2 dx$$
$$\leq CM\Big[\Big(\frac{\tilde{\varphi}(2r^*)}{\tilde{\varphi}(r^*/2)}\Big)^{2\lambda} \int_{\mathcal{B}(0,r^*)} |z|^2 dx + \Big(\frac{\tilde{\varphi}(2r^*)}{\tilde{\varphi}(5r^*/2)}\Big)^{2\lambda} \int_{\mathcal{B}(0,3r^*)} |z|^2 dx\Big]. \qquad (2.132)$$

Let

$$\lambda_2 = \frac{1}{2(\alpha+\beta)} \ln \Big[\int_{\mathcal{B}(0,3r^*)} |z|^2 dx \Big/ \int_{\mathcal{B}(0,r^*)} |z|^2 dx\Big], \qquad (2.133)$$

where α and β are given in (2.124).

If $\lambda_2 \geq \lambda_1$ (recall (2.127) for λ_1), then choosing $\lambda = \lambda_2$ in (2.132) yields

$$\int_{\mathcal{B}(0,2r^*)} |z|^2 dx \leq 2CM\Big(\int_{\mathcal{B}(0,r^*)} |z|^2 dx\Big)^{\frac{\alpha}{\alpha+\beta}} \Big(\int_{\mathcal{B}(0,3r^*)} |z|^2 dx\Big)^{\frac{\beta}{\alpha+\beta}}. \qquad (2.134)$$

If $\lambda_2 < \lambda_1$, then it follows from (2.133) that

$$\int_{\mathcal{B}(0,3r^*)} |z|^2 dx < e^{2(\alpha+\beta)\lambda_1} \int_{\mathcal{B}(0,r^*)} |z|^2 dx. \qquad (2.135)$$

By (2.135), and recalling (2.127) for the definition of λ_1, we have that

$$\int_{\mathscr{B}(0,2r^*)} |z|^2 dx \leq \int_{\mathscr{B}(0,3r^*)} |z|^2 dx \leq e^{2(\alpha+\beta)\lambda_1} \int_{\mathscr{B}(0,r^*)} |z|^2 dx$$
$$\leq e^{C(\alpha+\beta)M^{2/3}} \int_{\mathscr{B}(0,r^*)} |z|^2 dx. \tag{2.136}$$

Finally, combining (2.134) and (2.136), by making a suitable change of variable, we obtain the desired inequality (2.123).

Remark 2.17 The inequality (2.123) was established in [4]. It is a key tool to study the spectral localization of random Schrödinger operators on $\mathbb{R}^n$ (See [4] for more details).

Remark 2.18 Clearly, the choice of $\mathscr{U}$ and $\mathscr{K}$ in (2.125) is not unique, and hence, the corresponding numbers α and β in Theorem 2.7 are not unique, either.

Remark 2.19 Let s_1, s_2 and s_3 be positive real numbers such that $0 < s_1 < s_2 < s_3$. Theorem 2.7 can be extended to a more general case. In fact, similar to the proof of Theorem 2.7, one can show the following three-ball inequality:

$$|z|_{L^2(\mathscr{B}(0,s_2r))} \leq \left[CM + e^{C(\tilde{\alpha}+\tilde{\beta})M^{2/3}} \right] |z|_{L^2(\mathscr{B}(0,s_1r))}^{\frac{\tilde{\alpha}}{\tilde{\alpha}+\tilde{\beta}}} |z|_{L^2(\mathscr{B}(0,s_3r))}^{\frac{\tilde{\beta}}{\tilde{\alpha}+\tilde{\beta}}}, \tag{2.137}$$

where $0 < r < \frac{1}{s_3}\text{diam}\,(\Omega)$, and $\tilde{\alpha}$ and $\tilde{\beta}$ are given by

$$\tilde{\alpha} = \ln\left[\tilde{\varphi}(s_2r)/\tilde{\varphi}(\tilde{s}_1r) \right], \quad \tilde{\beta} = \ln\left[\tilde{\varphi}(\tilde{s}_3r)/\tilde{\varphi}(s_2r) \right] \tag{2.138}$$

with a suitable choice of $\tilde{s}_1 < s_1$, $\tilde{s}_3 < s_3$ and $\tilde{s}_1 \ll \tilde{s}_3$.

Remark 2.20 In (2.123), the index $2/3$ in $M^{2/3}$ is sharp, namely, this inequality fails if $2/3$ is replaced by α for any $\alpha < 2/3$ (See [4]).

As an application of Theorem 2.7, we give below two quantitative unique continuation results. The first one is as follows.

Theorem 2.8 *Let* $z \in L^\infty(\Omega; \mathbb{C})$ *be a solution to (2.122) with* $|c|_{L^\infty(\Omega)} \leq M$ *and* $1 \leq |z|_{L^\infty(\Omega)} \leq C_0$ *for some constants* $M \gg 1$ *and* $C_0 \geq 1$. *Then there exist two constants* $C_1 = C_1(n, C_0, M) > 0$ *and* $C_2 = C_2(n, C_0, M) > 0$ *such that*

$$m(r) = \sup_{|x| \leq r} |z| \geq C_1 r^{C_2 M^{2/3}}, \quad \forall\, r \in (0, 1]. \tag{2.139}$$

Proof Without loss of generality, we assume that $\mathscr{B}(0, 6) \subset \Omega$. Choosing $s_2r = 2$ and $s_3r = 6$ in (2.137), by (2.138), we have that

$$|z|_{L^2(\mathscr{B}(0,2))} \leq CM|z|_{L^2(\mathscr{B}(0,r_0))}^{\frac{\tilde{\alpha}}{\tilde{\alpha}+\tilde{\beta}}} |z|_{L^2(\mathscr{B}(0,6))}^{\frac{\tilde{\beta}}{\tilde{\alpha}+\tilde{\beta}}} + \left(\frac{\tilde{\varphi}(\tilde{s}_3r)}{\tilde{\varphi}(\tilde{s}_1r)} \right)^{CM^{2/3}} |z|_{L^2(\mathscr{B}(0,r_0))}, \tag{2.140}$$

where $\tilde{\alpha}$ and $\tilde{\beta}$ are given in (2.138), $r_0 \stackrel{\triangle}{=} s_1 r \ll 1$, $\tilde{s}_1 < s_1$, $\tilde{s}_1 \ll \tilde{s}_3$ and $\tilde{s}_3 r < 6$.

Recalling (2.22) the definition of w, without loss of generality, we assume that

$$\frac{\tilde{\varphi}(\tilde{s}_3 r)}{\tilde{\varphi}(\tilde{s}_1 r)} = O\left(\frac{1}{r_0}\right). \tag{2.141}$$

Choose a cut-off function $\zeta \in C_0^2(\mathscr{B}(0, 1 + 2/n); [0, 1])$ such that $\zeta(x) = 1$ on $\mathscr{B}(0, 1)$. Multiplying (2.122) by $\zeta^2 \bar{z}$, we get that

$$\sum_{j,k=1}^n \left[-\left(h^{jk} z_{x_j} \zeta^2 \bar{z}\right)_{x_k} + h_{x_k}^{jk} z_{x_j} \zeta^2 \bar{z} + h^{jk} z_{x_j} \zeta \zeta_{x_k} \bar{z} + \zeta^2 h^{jk} z_{x_j} \bar{z}_{x_k} \right] + a\zeta^2 |z|^2 = 0.$$

Hence,

$$
\begin{aligned}
h_0 \zeta^2 |\nabla z|^2 &\leq \zeta^2 \sum_{j,k=1}^n h^{jk} z_{x_j} \bar{z}_{x_k} \\
&\leq \sum_{j,k=1}^n (h^{jk} z_{x_j} \zeta^2 \bar{z})_{x_k} + \varepsilon \zeta^2 |\nabla z|^2 + \frac{C_1}{\varepsilon} |z|^2 + |c|_{L^\infty(\mathscr{B}(0,1+2/n))} \zeta^2 |z|^2.
\end{aligned}
\tag{2.142}
$$

By integrating (2.142) on Ω, we find that there is a constant $C_1 > 0$ such that

$$\int_{\mathscr{B}(0,1)} (|z|^2 + |\nabla z|^2) dx \leq C_1 (M + 1) |z|_{L^2(\mathscr{B}(0,1+2/n))}^2. \tag{2.143}$$

By iteration, we can get that there is a constant $C_n > 0$ such that

$$|z|_{H^n(\mathscr{B}(0,1))}^2 \leq C_n (M^n + 1) |z|_{L^2(\mathscr{B}(0,2))}^2. \tag{2.144}$$

This, together with Sobolev's embedding theorem, implies that

$$|z|_{L^\infty(\mathscr{B}(0,1))} \leq C_n (M^{n/2} + 1) |z|_{L^2(\mathscr{B}(0,2))} \leq C_n M^{n/2} |z|_{L^2(\mathscr{B}(0,2))}. \tag{2.145}$$

Combining (2.140) and (2.141)–(2.145), we have

$$|z|_{L^\infty(\mathscr{B}(0,1))} \leq J_1 + J_2,$$

where

$$J_1 = C C_n M^{n/2+1} |z|_{L^2(\mathscr{B}(0,r_0))}^{\frac{\tilde{\alpha}}{\tilde{\alpha}+\tilde{\beta}}} |z|_{L^2(\mathscr{B}(0,6))}^{\frac{\tilde{\beta}}{\tilde{\alpha}+\tilde{\beta}}},$$

and

$$J_2 = C_n M^{n/2} \left(\frac{C}{r_0}\right)^{CM^{2/3}} |z|_{L^2(\mathscr{B}(0,r_0))}.$$

Let us consider two different cases.

Case 1. $J_1 \leq J_2$. By (2.145), a short calculation shows that

$$
\begin{aligned}
|z|_{L^\infty(\mathscr{B}(0,1))} &\leq 2C_n M^{n/2}\left(\frac{C}{r_0}\right)^{CM^{2/3}} |z|_{L^2(\mathscr{B}(0,r_0))} \\
&\leq 2C_n M^{n/2}\left(\frac{C}{r_0}\right)^{CM^{2/3}} \sup_{x\in\mathscr{B}(0,r_0)} |z(x)|,
\end{aligned}
$$

which gives the desired lower bound.

Case 2. $J_1 \geq J_2$. Recalling that $|z|_{L^\infty(\mathscr{B}(0,1))} \geq 1$, it holds that

$$
1 \leq |z|_{L^\infty(\mathscr{B}(0,1))} \leq 2CC_n M^{n/2+1}|z|_{L^2(\mathscr{B}(0,r_0))}^{\frac{\tilde{\alpha}}{\tilde{\alpha}+\tilde{\beta}}} |z|_{L^2(\mathscr{B}(0,6))}^{\frac{\tilde{\beta}}{\tilde{\alpha}+\tilde{\beta}}}.
$$

Raising both sides to $\frac{\tilde{\alpha}+\tilde{\beta}}{\tilde{\alpha}}$ and using the bound $|z|_{L^\infty(\mathscr{B}(0,6))} \leq C_0$, we obtain

$$
\begin{aligned}
1 &\leq \left(2CC_n M^{n/2+1}\right)^{\frac{\tilde{\alpha}+\tilde{\beta}}{\tilde{\alpha}}} |z|_{L^\infty(\mathscr{B}(0,r_0))} C_0^{\frac{\tilde{\beta}}{\tilde{\alpha}}} \\
&\leq \left(2CC_0C_n M^{n/2+1}\right)^{\frac{\tilde{\alpha}+\tilde{\beta}}{\tilde{\alpha}}} |z|_{L^\infty(\mathscr{B}(0,r_0))}.
\end{aligned} \tag{2.146}
$$

Recalling (2.138) for $\tilde{\alpha}$ and $\tilde{\beta}$, and noting that $\tilde{k} < k \ll 1$, by the definition of w and (2.141), we find that

$$
\frac{\tilde{\alpha}+\tilde{\beta}}{\tilde{\alpha}} = \frac{1}{\tilde{\alpha}}\ln\left(\frac{w(\tilde{m}r)}{w(\tilde{k}r)}\right) \leq \frac{C}{\ln(2/\tilde{k})}\ln\frac{C}{r_0} \leq C\ln\frac{1}{r_0}.
$$

Hence, the right hand side of (2.146) is bounded by $r_0^{-C\ln(CC_0C_n M^{n/2+1})}$. The desired result follows.

Remark 2.21 In (2.139), the index 2/3 of M is optimal (e.g., [45]).

The second quantitative unique continuation result is stated below.

Theorem 2.9 *Suppose that $z \in L^\infty(\mathbb{R}^n; \mathbb{C})$ is a solution to*

$$
\sum_{j,k=1}^n (h^{jk}z_{x_j})_{x_k} + \tilde{a}z = 0 \quad \text{in } \mathbb{R}^n
$$

with $|\tilde{a}|_{L^\infty(\mathbb{R}^n)} \leq 1$, $|z|_{L^\infty(\mathbb{R}^n)} \leq C_0$ for a given constant $C_0 > 0$ and $z(0) = 1$. Then there exist a constant $C > 0$ such that

$$
M(R) \equiv \inf_{|x_0|=R} \sup_{x\in\mathscr{B}(x_0,1)} |z(x)| \geq Ce^{-R^{4/3}\ln R}, \quad \forall\, R > 0. \tag{2.147}
$$

Proof Fix an x_0 such that $|x_0| = R$ and that

$$M(R) = \inf_{|x_0|=R} \sup_{x \in \mathscr{B}(x_0,1)} |z(x)| = \sup_{\mathscr{B}(x_0,1)} |z(x)|.$$

Set

$$z_R(x) = z(Rx + x_0) = z(R(x + x_0/R)).$$

Then, it is easy to get that

$$|z_R|_{L^\infty(\mathbb{R}^n)} \le C_0, \quad \left| \sum_{j,k=1}^{n} (h^{jk} \partial_{x_j} z_R)_{x_k} \right| \le R^2 |z_R|.$$

Let $\tilde{x}_0 = -x_0/R$. Then $|\tilde{x}_0| = 1$ and $z_R(\tilde{x}_0) = z(0) = 1$. Thus $|z_R|_{L^\infty(\mathscr{B}(0,1))} \ge 1$ and $\sup_{x \in \mathscr{B}(x_0,1)} |z(x)| = \sup_{y \in \mathscr{B}(0,r_0)} |z_R(y)|$, where $r_0 = 1/R$.

Now, using Theorem 2.8 with $M = R^2$, we have that

$$M(R) = \sup_{y \in \mathscr{B}(0,r_0)} |z_R(y)| \ge Cr_0^{M^{2/3}} = C(1/R)^{R^{4/3}} = Ce^{-CR^{4/3} \ln R}.$$

Remark 2.22 Due to the example in [45], one can show that the index $4/3$ in (2.147) is sharp.

2.6   Further Comments

In this chapter, we present some typical Carleman estimates for elliptic equations and their applications. The main tool is the pointwise identity in Lemma 2.1. There are some important results that are not covered here. Some of them are as follows (See also the very interesting recent book [30] for more material):

- **The L^p ($p \in [1, \infty)$) Carleman estimate for elliptic equations** (e.g. [22, 25, 48]). It can be applied to establish SUCP for solutions to elliptic equations with L^p-integrable lower-order coefficients. This is very important in the study of the spectrum of Schrödinger operators (e.g. [23, 47]). In [23], SUCP was established for (2.114) when $b \in L_{loc}^{\frac{n}{2}}(\Omega)$. Their result is sharp. Indeed, SUCP fails for (2.114) when $b \in L^p(\Omega)$ for $p < \frac{n}{2}$. An example is given as follows:

    Let $y(x) = \exp(-\ln \frac{1}{|x|})^{1+\varepsilon}$, $\varepsilon > 0$. Then y vanishes at 0 of infinity order, while

    $$V(x) = -\frac{\Delta y(x)}{y(x)} \sim \frac{(\ln \frac{1}{|x|})^{2\varepsilon}}{|x|^2} \in L_{loc}^p(\mathbb{R}^n), \quad \forall \, p < \frac{n}{2}.$$

Hence SUCP cannot hold for $p < n/2$.

In [22, 23, 25, 47, 48], many tools in Harmonic Analysis, such as oscillatory integrals, restriction theorems for the Fourier transform, complex interpolation and the uncertainty principle, were employed to prove the L^p Carleman estimate. A different method was introduced in [27], which combines the L^2 Carleman estimate and dispersive estimate to derive the L^p Carleman estimate. Nevertheless, as far as we know, the latter cannot give the sharp estimate. It would be quite interesting to prove the results in [23, 47] by the method provided in this chapter but so far we do not know how to do it.

- **Carleman estimate for elliptic equations with a limiting/degenerate Carleman weight** (e.g. [10, 21, 24]). These estimates play a key role in solving some inverse problems of elliptic equations (e.g. [10, 21, 24]). They are established by some tools in Semiclassical Analysis. Whether they can be obtained by methods in this chapter is an interesting open problem.

- **Carleman estimate for elliptic equations with coefficients which has jumps at an interface** (e.g. [2, 8]). The main results in [2] were obtained by semiclassical analysis techniques. Then in [8], a more elementary method (based on the straightforward Fourier transform) was provided. We believe that the result in [8] can be obtained by the method in this chapter. The key point is to find a suitable weight function to handle the jumps. Due to the limitation of length, we will not discuss the details here.

- **Carleman estimates for elliptic operators on Riemann manifolds** (e.g. [1]). Such sort of estimates are very useful in some problems in Geometry. For example, it can be used to prove the finiteness theorems for cohomology with coefficients on a locally free sheaf on a complex manifold.

- **Carleman estimates for elliptic equations with nonhomogeneous Dirichlet boundary and nonhomogeneous terms of low regularity** (e.g. [20]). The main motivation for this sort of Carleman estimates is to obtain sharp estimate for the pressure in linearized Navier-Stokes equations, We do not present the details here and refer the readers to [3, 11, 16, 20].

- **Carleman estimate for elliptic systems** (e.g. [35, 36]). One can employ the method in this chapter to obtain the results in [35, 36]. Due to the limitation of space, we will not give the details here.

References

1. Andreotti, A., Vesentini, E.: Carleman estimates for the Laplace-Beltrami equation on complex manifolds. Inst. Hautes études Sci. Publ. Math. **25**, 81–130 (1965)
2. Bellassoued, M., Le Rousseau, J.: Carleman estimates for elliptic operators with complex coefficients. Part II: transmission problems. J. Math. Pures Appl. **115**, 127–186 (2018)
3. Boulakia, M., Guerrero, S.: Local null controllability of a fluid-solid interaction problem in dimension 3. J. Eur. Math. Soc. **15**, 825–856 (2013)

4. Bourgain, J., Kenig, C.: On localization in the continuous Anderson-Bernoulli model in higher dimension. Invent. Math. **161**, 389–426 (2005)
5. Burq, N.: Décroissance de l'énergie locale de l'équation des ondes pour le probléme extérieur et absence de résonance au voisinagage du réel. Acta Math. **180**, 1–29 (1998)
6. Carleman, T.: Sur un problème d'unicité pour les systèmes d'équations aux dérivées partielles à deux variables indépendantes. Ark. Mat. Astr. Fys. 26 B. **17**, 1–9 (1939)
7. Choulli, M.: Applications of Elliptic Carleman Inequalities to Cauchy and Inverse Problems. Springer Briefs in Mathematics. BCAM Springer Briefs, Springer, Cham (2016)
8. Di Cristo, M., Francini, E., Lin, C., Vessella, S., Wang, J.: Carleman estimate for second order elliptic equations with Lipschitz leading coefficients and jumps at an interface. J. Math. Pures Appl. **108**, 163–206 (2017)
9. Donnelly, H., Fefferman, C.: Nodal sets of eigenfunctions on Riemannian manifolds. Invent. Math. **93**, 161–183 (1988)
10. Dos Santos Ferreira, D., Kenig, C.E., Salo, M., Uhlmann, G.: Limiting Carleman weights and anisotropic inverse problems. Invent. Math. **178**, 119–171 (2009)
11. Fernández-Cara, E., Guerrero, S., Imanuvilov, OYu., Puel, J.-P.: Local exact controllability of the Navier-Stokes system. J. Math. Pures Appl. **83**, 1501–1542 (2004)
12. Fu, X.: Logarithmic decay of hyperbolic equations with arbitrary small boundary damping. Commun. Partial Differ. Equ. **34**, 957–975 (2009)
13. Fu, X.: Longtime behavior of the hyperbolic equations with an arbitrary internal damping. Z. Angew. Math. Phys. **62**, 667–680 (2011)
14. Fu, X.: Sharp decay rates for the weakly coupled hyperbolic system with one internal damping. SIAM J. Control Optim. **50**, 1643–1660 (2012)
15. Fursikov, A.V., Imanuvilov, O.Y.: Controllability of Evolution Equations. Lecture Notes Series, vol. 34. Seoul National University, Seoul, Korea (1996)
16. Guerrero, S.: Local exact controllability to the trajectories of the Boussinesq system. Ann. Inst. H. Poincaré Anal. Non Linéaire **23**, 29–61 (2006)
17. Hörmander, L.: Linear Partial Differential Operators. Springer, Berlin-Göttingen-Heidelberg (1963)
18. Hörmander, L.: Uniqueness theorems for second order elliptic differential equations. Commun. Partial Differ. Equ. **8**, 21–64 (1983)
19. Hörmander, L.: The Analysis of Linear Partial Differential Operators. IV. Fourier Integral Operators. Springer, Berlin (1985)
20. Imanuvilov, O.Y., Puel, J-P.: Global Carleman estimates for weak solutions of elliptic nonhomogeneous Dirichlet problems. Int. Math. Res. Not. **16**, 883–913 (2003)
21. Imanuvilov, OYu., Uhlmann, G., Yamamoto, M.: The Calderón problem with partial data in two dimensions. J. Am. Math. Soc. **23**, 655–691 (2010)
22. Jerison, D., Lebeau, G.: Nodal sets of sums of eigenfunctions. Harmonic Analysis and Partial Differential Equations (Chicago, IL, 1996). Chicago Lectures in Mathematics, pp. 223–239. University of Chicago Press, Chicago, IL (1999)
23. Jerison, D., Kenig, C.E.: Unique continuation and absence of positive eigenvalues for Schrödinger operators. Ann. Math. **121**, 463–494 (1985)
24. Kenig, C.E., Sjöstrand, J., Uhlmann, G.: The Calderón problem with partial data. Ann. Math. **165**, 567–591 (2007)
25. Koch, H., Tataru, D.: Carleman estimates and unique continuation for second order elliptic equations with nonsmooth coefficients. Comm. Pure Appl. Math. **54**, 339–360 (2001)
26. Koch, H., Tataru, D.: Sharp counterexamples in unique continuation for second order elliptic equations. J. Reine Angew. Math. **542**, 133–146 (2002)
27. Koch, H., Tataru, D.: Dispersive estimates for principally normal pseudodifferential operators. Comm. Pure Appl. Math. **58**, 217–284 (2005)
28. Le Rousseau, J., Robbiano, L.: Spectral inequality and resolvent estimate for the bi-Laplace operator. J. Eur. Math. Soc. to appear
29. Le Rousseau, J., Léautaud, M., Robbiano, L.: Controllability of a parabolic system with a diffuse interface. J. Eur. Math. Soc. **15**, 1485–1574 (2013)

30. Lebeau, G., Le Rousseau, J., Robbiano, L.: Elliptic Carleman Estimates and Applications to Stabilization and Controllability. Volume 1: Dirichlet boundary condition on Euclidean space. Book in preparation

31. Lebeau, G.: Equation des ondes amorties, in Algebraic and Geometric Methods in Mathematical Physics (Kaciveli, 1993), pp. 73–109. Kluwer Academic Publishers, Dordrecht (1996)

32. Lebeau, G., Robbiano, L.: Contrôle exact de l'équation de la chaleur. Commun. Partial Differ. Equ. **20**, 335–356 (1995)

33. Lebeau, G., Robbiano, L.: Stabilisation de l'équation des ondes par le bord. Duke Math. J. **86**, 465–491 (1997)

34. Lebeau, G., Zuazua, E.: Null-controllability of a system of linear thermoelasticity. Arch. Ration. Mech. Anal. **141**, 297–329 (1998)

35. Lin, C., Wang, J.-N.: Strong unique continuation for the Lamé system with Lipschitz coefficients. Math. Ann. **331**, 611–629 (2005)

36. Lin, C., Nakamura, G., Wang, J.-N.: Optimal three-ball inequalities and quantitative uniqueness for the Lamé system with Lipschitz coefficients. Duke Math. J. **155**, 189–204 (2010)

37. Liu, X.: Controllability of some coupled stochastic parabolic systems with fractional order spatial differential operators by one control in the drift. SIAM J. Control Optim. **52**, 836–860 (2014)

38. Liu, Z., Rao, B.: Characterization of polynomial decay rate for the solution of linear evolution equation. Z. Angew. Math. Phys. **56**, 630–644 (2005)

39. López, A., Zhang, X., Zuazua, E.: Null controllality of the heat equation as singular limit of the exact controllability of dissipative wave equations. J. Math. Pures Appl. **79**, 741–808 (2000)

40. Lü, Q.: A lower bound on local energy of partial sum of eigenfunctions for Laplace-Beltrami operators. ESAIM: Control Optim. Calc. Var. **19**, 255–273 (2013)

41. Lü, Q.: Some results on the controllability of forward stochastic heat equations with control on the drift. J. Funct. Anal. **260**, 832–851 (2011)

42. Lü, Q., Wang, G.: On the existence of time optimal controls with constraints of the rectangular type for heat equations. SIAM J. Control Optim. **49**, 1124–1149 (2011)

43. Lü, Q., Yin, Z.: The L^∞-null controllability of parabolic equation with equivalued surface boundary conditions. Asymptot. Anal. **83**, 355–378 (2013)

44. Lü, Q., Zuazua, E.: Robust null controllability for heat equations with unknown switching control mode. Discret. Contin. Dyn. Syst. **34**, 4183–4210 (2014)

45. Meshkov, V.Z.: On the possible rate of decay at infinity of solutions of second order partial differential equations. Math. USSR Sbornik. **72**, 343–360 (1992)

46. Pliś, A.: On non-uniqueness in Cauchy problem for an elliptic second order differential operator. Bull. Acad. Polon. Sci. Sér. Sci. Math. Astronom. Phys. **11**, 95–100 (1963)

47. Simon, B.: Schrödinger semigroups. Bull. Am. Math. Soc. **7**, 447–526 (1982)

48. Sogge, C.D.: Oscillatory integrals, Carleman inequalities and unique continuation for second order elliptic differential equations. J. Am. Soc. Math. **2**, 491–516 (1989)

49. Wang, G.: L^∞-null controllability for the heat equation and its consequences for the time optimal control problem. SIAM J. Control Optim. **48**, 1701–1720 (2008)

Chapter 3
Carleman Estimates for Second Order Parabolic Operators and Applications, a Unified Approach

Abstract In this chapter, we establish three Carleman estimates with different weight functions for second order parabolic operators. The first one is Theorem 3.1, which is used to obtain controllability/observability results for parabolic equations in Sect. 3.2. The second one is Theorem 3.2, via which, we solve an inverse parabolic problem in Sect. 3.3. The third one is Theorem 3.3, and it yields the SUCP of parabolic equations in Sect. 3.4.

Keywords Carleman estimate · Second order parabolic operator · Null controllability · Strong unique continuation · Three cylinders inequality

To begin with, we introduce the following condition:

Condition 3.1 *Let $p^{jk}(\cdot) \in C^1(\overline{Q}; \mathbb{R})$ $(j, k = 1, \ldots, n)$ be fixed functions satisfying*

$$p^{jk}(t, x) = p^{kj}(t, x), \quad \forall (t, x) \in \overline{Q}, \ j, k = 1, 2, \ldots, n, \tag{3.1}$$

and for some constant $p_0 > 0$,

$$\sum_{j,k=1}^{n} p^{jk}(t, x)\xi^j\overline{\xi}^k \geq p_0|\xi|^2, \quad \forall (t, x, \xi^1, \ldots, \xi^n) \in \overline{Q} \times \mathbb{C}^n. \tag{3.2}$$

In this chapter, we denote by $C = C(\Omega, \omega, n, (p^{jk})_{1 \leq j,k \leq n})$ a generic positive constant, which may change from line to line (unless otherwise stated).

3.1 Carleman Estimates for Second Order Parabolic Operators

We first establish a pointwise inequality for second order parabolic operators.

Recalling the definition of $\mathscr{P}$ in (1.11) and choosing

$$\alpha = -1, \quad \beta = 0, \quad m = n, \quad a^{jk} = p^{jk}, \quad j, k = 1, 2, \ldots, n,$$

by Theorem 1.1, and using Hölder's inequality, we have the following result.

Corollary 3.1 *Let $z \in C^2(\mathbb{R}^{1+n}; \mathbb{R})$, $\ell \in C^2(\mathbb{R}^{1+n}; \mathbb{R})$, $\Psi \in C^1(\mathbb{R}^{1+n}; \mathbb{R})$ and $\Phi \in C(\mathbb{R}^{1+n}; \mathbb{R})$. Then (see (1.13) for θ and v)*

$$\theta^2 \left| z_t - \sum_{j,k=1}^n (p^{jk} z_{x_j})_{x_k} \right|^2 + \left[(\ell_t + A)|v|^2 - \sum_{j,k=1}^n p^{jk} v_{x_j} v_{x_k} \right]_t + \operatorname{div} V$$

$$\geq |I_1 + \Phi v|^2 + 2 \sum_{j,k=1}^n c^{jk} v_{x_j} v_{x_k} - 2 \sum_{j,k=1}^n p^{jk} \Psi_{x_j} v_{x_k} v + B|v|^2, \tag{3.3}$$

where

$$\begin{cases} A = \displaystyle\sum_{j,k=1}^n p^{jk} \ell_{x_j} \ell_{x_k} - \sum_{j,k=1}^n (p^{jk} \ell_{x_j})_{x_k} - \Psi - \Phi, \\[4mm] I_1 = \ell_t v + \displaystyle\sum_{j,k=1}^n (p^{jk} v_{x_j})_{x_k} + Av, \end{cases} \tag{3.4}$$

and

$$\begin{cases} V = [V^1, V^2, \ldots, V^n], \\[2mm] V^k = \displaystyle\sum_{j=1}^n \Big[2p^{jk} v_{x_j} v_t + 2 \sum_{j',k'=1}^n \Big(2p^{jk'} p^{j'k} - p^{jk} p^{j'k'} \Big) \ell_{x_j} v_{x_{j'}} v_{x_{k'}} \\[4mm] \qquad\qquad - 2\Psi p^{jk} v_{x_j} v + p^{jk} \big(2A\ell_{x_j} + 2\ell_{x_j} \ell_t \big) |v|^2 \Big], \\[4mm] c^{jk} = \displaystyle\sum_{j',k'=1}^n \big[2\big(p^{j'k} \ell_{x_{j'}} \big)_{x_{k'}} p^{jk'} - \big(p^{jk} p^{j'k'} \ell_{x_{j'}} \big)_{x_{k'}} \big] - \frac{1}{2} p_t^{jk} - p^{jk} \Psi, \\[4mm] B = 2A\Psi + 2 \displaystyle\sum_{j,k=1}^n \big(p^{jk} \ell_{x_j} A \big)_{x_k} + 2 \sum_{j,k=1}^n \big(p^{jk} \ell_{x_j} \ell_t \big)_{x_k} + 2\Psi \ell_t + \ell_{tt} + A_t - \Phi^2. \end{cases} \tag{3.5}$$

Remark 3.1 From the definition of I_1 in (3.4), it is easy to see that, Corollary 3.1 simplifies a similar pointwise inequality in [31].

Based on Corollary 3.1, one can obtain some global Carleman estimates for the parabolic operator $\mathscr{P}_p \triangleq \partial_t - \displaystyle\sum_{j,k=1}^n \partial_{x_k} (p^{jk} \partial_{x_j})$, which are very useful in the study of control problems, inverse problems and SUCP for parabolic equations, etc. To this aim, we first introduce the weight functions. For any nonnegative and nonzero function $\psi \in C^2(\overline{\Omega})$, and any real numbers $\lambda > 1$ and $\mu > 1$, put

$$\varphi(t, x) = \frac{e^{\mu \psi(x)}}{t(T - t)}, \quad \tilde{\alpha}(t, x) = \frac{e^{\mu \psi(x)} - e^{2\mu |\psi|_{C(\overline{\Omega})}}}{t(T - t)}, \quad \theta = e^\ell, \quad \ell = \lambda \tilde{\alpha}. \tag{3.6}$$

For $j, k = 1, \ldots, n$, it is easy to see that

$$\ell_t = \lambda \tilde{\alpha}_t, \quad \ell_{x_j} = \lambda \mu \varphi \psi_{x_j}, \quad \ell_{x_j x_k} = \lambda \mu^2 \varphi \psi_{x_j} \psi_{x_k} + \lambda \mu \varphi \psi_{x_j x_k} \tag{3.7}$$

and

$$|\tilde{\alpha}_t| \le C\varphi^2, \quad |\tilde{\alpha}_{tt}| \le C\varphi^3. \tag{3.8}$$

Remark 3.2 The above choice of the "singular" weight function ℓ (at $t = 0$) so that $e^{\ell(0,x)} \equiv 0$ will be used to drop the terms concerning the initial data of solutions to parabolic equations. To the authors' knowledge, such a choice of the weight function (in Carleman estimates for parabolic equations) comes from [21].

We have the following global Carleman estimate.

Theorem 3.1 *Let φ be given in (3.6) and ψ be given by Lemma 2.4. Then there exists $\mu_0 > 0$ such that for all $\mu \ge \mu_0$, one can find two constants $C > 0$ and $\lambda_1 = \lambda_1(\mu)$ so that for all $z \in C([0, T]; L^2(\Omega)) \cap L^2(0, T; H_0^1(\Omega))$ solving*

$$z_t - \sum_{j,k=1}^n (p^{jk} z_{x_j})_{x_k} = g \in L^2(Q) \tag{3.9}$$

and all $\lambda \ge \lambda_1$, it holds that

$$\int_Q (\lambda \varphi)^{-1} \theta^2 \left(|z_t|^2 + \left| \sum_{j,k=1}^n (p^{jk} z_{x_j})_{x_k} \right|^2 \right) dxdt$$
$$+ \lambda^3 \mu^4 \int_Q \theta^2 \varphi^3 |z|^2 dxdt + \lambda \mu^2 \int_Q \theta^2 \varphi |\nabla z|^2 dxdt \tag{3.10}$$
$$\le C \left(|\theta g|_{L^2(Q)}^2 + \lambda^3 \mu^4 \int_0^T \int_\omega \varphi^3 \theta^2 |z|^2 dxdt \right).$$

Proof We divide the proof into three steps.

   Step 1. First, integrating (3.3) on Q, and noting that $\theta(0, \cdot) = \theta(T, \cdot) \equiv 0$, by (3.9), we obtain that

$$\int_Q |I_1 + \Phi v|^2 dxdt + \int_Q \left(2 \sum_{j,k=1}^n c^{jk} v_{x_k} v_{x_j} - 2 \sum_{j,k=1}^n p^{jk} \Psi_{x_k} v_{x_j} v + B|v|^2 \right) dxdt$$
$$\le |\theta g|_{L^2(Q)}^2 + \int_\Sigma V \cdot v dxdt. \tag{3.11}$$

We choose

$$\Psi = -2\lambda \mu^2 \varphi \sum_{j,k=1}^n p^{jk} \psi_{x_j} \psi_{x_k}, \quad \Phi = -\Psi - \sum_{j,k=1}^n (p^{jk} \ell_{x_j})_{x_k}. \tag{3.12}$$

By (3.7) and (3.12), recalling (3.4) for the definition of A, it is easy to find that

$$A = \sum_{j,k=1}^{n} p^{jk} \ell_{x_j} \ell_{x_k} = \lambda^2 \mu^2 \varphi^2 \sum_{j,k=1}^{n} p^{jk} \psi_{x_j} \psi_{x_k}, \quad \varPhi = \lambda \varphi O(\mu^2). \tag{3.13}$$

Recalling (3.5) and (3.6) respectively for the definitions of B and φ, by (3.13) and (3.8), we have

$$\begin{aligned}
B &= 2A\Psi + 2 \sum_{j,k=1}^{n} (p^{jk} \ell_{x_j} A)_{x_k} + 2 \sum_{j,k=1}^{n} (p^{jk} \ell_{x_j} \ell_t)_{x_k} + 2\Psi \ell_t + \ell_{tt} + A_t - \varPhi^2 \\
&= 2\lambda^3 \mu^4 \varphi^3 \sum_{j,k=1}^{n} \left(p^{jk} \psi_{x_j} \psi_{x_k} \right)^2 + \lambda^3 \varphi^3 O(\mu^3) + \lambda^2 \varphi^2 O(\mu^4).
\end{aligned} \tag{3.14}$$

Further,

$$\begin{aligned}
&\sum_{j,k=1}^{n} \left(c^{jk} v_{x_j} v_{x_k} - p^{jk} \Psi_{x_j} v_{x_k} v \right) \\
&\geq \lambda \mu^2 \varphi \left(\sum_{j',k'=1}^{n} p^{j'k'} \psi_{x_{j'}} \psi_{x_{k'}} \right) \left(\sum_{j,k=1}^{n} p^{jk} v_{x_j} v_{x_k} \right) \\
&\quad - C\lambda \mu \varphi \left(\mu^2 |v| |\nabla v| + |\nabla v|^2 \right) \\
&\geq \lambda \mu^2 \varphi p_0^2 |\nabla \psi|^2 |\nabla v|^2 - C\lambda^2 \mu^4 \varphi^2 |v|^2 - C\mu^2 |\nabla v|^2 - C\lambda \mu \varphi |\nabla v|^2.
\end{aligned} \tag{3.15}$$

Next, recalling the definition of V in (3.5), and noting that $v_{x_j} = \frac{\partial v}{\partial \nu} \nu^j$ (which follows from $v = \theta z$ and $v|_\Sigma = 0$), $\psi|_\Gamma = 0$, $\frac{\partial \psi}{\partial \nu}\big|_\Gamma \leq 0$ and $z|_\Sigma = 0$, we obtain that

$$\begin{aligned}
\int_\Sigma V \cdot \nu dx dt &= \int_\Sigma \sum_{j,k,j',k'=1}^{n} \Big[2p^{jk} v_{x_j} v_t + 2 \left(2p^{jk} p^{j'k} - p^{jk} p^{j'k'} \right) \ell_{x_j} v_{x_{j'}} v_{x_{k'}} \\
&\qquad - 2\Psi p^{jk} v_{x_j} v + p^{jk} (2A\ell_{x_j} + 2\ell_{x_j} \ell_t) |v|^2 \Big] \nu^k dx dt \\
&= 2\lambda \mu \int_\Sigma \varphi \frac{\partial \psi}{\partial \nu} \Big| \frac{\partial v}{\partial \nu} \Big|^2 \left(\sum_{j,k=1}^{n} p^{jk} \nu^j \nu^k \right)^2 dx dt \leq 0.
\end{aligned} \tag{3.16}$$

Step 2. It follows from (3.11), (3.14) and (3.16) that

$$\begin{aligned}
&\int_Q |I_1 + \varPhi v|^2 dx dt + \lambda \mu^2 \int_Q \varphi |\nabla \psi|^2 |\nabla v|^2 dx dt + \lambda^3 \mu^4 \int_Q \varphi^3 |\nabla \psi|^4 |v|^2 dx dt \\
&\leq C \Big[\int_Q \theta^2 |g|^2 dx dt + \mu \int_Q (\lambda \varphi + \mu) |\nabla v|^2 dx dt \\
&\qquad + \lambda^2 \mu^2 \int_Q \varphi^2 \left(\lambda \mu \varphi + \mu^2 \right) |v|^2 dx dt \Big].
\end{aligned} \tag{3.17}$$

Further, by $\min\limits_{x\in\Omega\setminus\omega_0}|\nabla\psi| > 0$, we get that

$$
\begin{aligned}
&\lambda\mu^2\int_Q \varphi|\nabla\psi|^2|\nabla v|^2 dxdt + \lambda^3\mu^4\int_Q \varphi^3|\nabla\psi|^4|v|^2 dxdt \\
&= \int_0^T \Big(\int_{\Omega\setminus\omega_0} + \int_{\omega_0}\Big)\big(\lambda\mu^2\varphi|\nabla\psi|^2|\nabla v|^2 + \lambda^3\mu^4\varphi^3|\nabla\psi|^4|v|^2\big)dxdt \\
&\geq c_0\lambda\mu^2\int_0^T\int_{\Omega\setminus\omega_0}\varphi\big(|\nabla v|^2 + \lambda^2\mu^2\varphi^2|v|^2\big)dxdt \\
&\quad - C\lambda\mu^2\int_0^T\int_{\omega_0}\varphi\big(|\nabla v|^2 + \lambda^2\mu^2\varphi^2|v|^2\big)dxdt,
\end{aligned}
\tag{3.18}
$$

where $c_0 = \min\big(\min\limits_{x\in\Omega\setminus\omega_0}|\nabla\psi|^2, \min\limits_{x\in\Omega\setminus\omega_0}|\nabla\psi|^4\big) > 0$.

By

$$
z_{x_j} = \theta^{-1}(v_{x_j} - \ell_{x_j}v) = \theta^{-1}(v_{x_j} - \lambda\mu\varphi\psi_{x_j}v)
$$

and

$$
v_{x_j} = \theta(z_{x_j} + \ell_{x_j}z) = \theta(z_{x_j} + \lambda\mu\varphi\psi_{x_j}z),
$$

we obtain that

$$
\frac{1}{C}\theta^2\big(|\nabla z|^2 + \lambda^2\mu^2\varphi^2|z|^2\big) \leq |\nabla v|^2 + \lambda^2\mu^2\varphi^2|v|^2 \leq C\theta^2\big(|\nabla z|^2 + \lambda^2\mu^2\varphi^2|z|^2\big).
$$

Consequently, it follows from (3.18) that

$$
\begin{aligned}
&\lambda\mu^2\int_Q \varphi\theta^2\big(|\nabla z|^2 + \lambda^2\mu^2\varphi^2|z|^2\big)dxdt \\
&= \lambda\mu^2\int_0^T\Big(\int_{\Omega\setminus\omega_0} + \int_{\omega_0}\Big)\varphi\theta^2\big(|\nabla z|^2 + \lambda^2\mu^2\varphi^2|z|^2\big)dxdt \\
&\leq C\Big[\int_Q \lambda\mu^2\varphi|\nabla\psi|^2|\nabla v|^2 dxdt + \int_Q \lambda^3\mu^4\varphi^3|\nabla\psi|^4|v|^2 dxdt \\
&\quad + \lambda\mu^2\int_0^T\int_{\omega_0}\varphi\theta^2\big(|\nabla z|^2 + \lambda^2\mu^2\varphi^2|z|^2\big)dxdt\Big].
\end{aligned}
\tag{3.19}
$$

Combining (3.17) and (3.19), we conclude that there is a $\mu_1 > 0$ such that for all $\mu \geq \mu_1$, one can find a positive constant $\lambda_1 = \lambda_1(\mu)$ so that for any $\lambda \geq \lambda_1$, it holds that

$$
\int_Q |I_1 + \Phi v|^2 dxdt + \lambda\mu^2 \int_Q \varphi\theta^2 \big(|\nabla z|^2 + \lambda^2\mu^2\varphi^2|z|^2\big)dxdt
$$
$$
\le C\Big[\int_Q \theta^2|g|^2 dxdt + \lambda\mu^2 \int_0^T \int_{\omega_0} \varphi\theta^2\big(|\nabla z|^2 + \lambda^2\mu^2\varphi^2|z|^2\big)dxdt\Big]. \tag{3.20}
$$

Let $\zeta \in C_0^\infty([0,1]; \omega)$ be such that $\zeta \equiv 1$ on ω_0. By

$$
\big(\zeta^2\varphi\theta^2|z|^2\big)_t = \zeta^2|z|^2(\varphi\theta^2)_t + 2\zeta^2\varphi\theta^2 zz_t,
$$

and noting $\theta(0,\cdot) = \theta(T,\cdot) \equiv 0$, we find

$$
\begin{aligned}
0 &= \int_0^T \int_{\omega_0} \zeta^2\Big[|z|^2(\varphi\theta^2)_t + 2\varphi\theta^2 zz_t\Big]dxdt \\
&= \int_0^T \int_{\omega_0} \zeta^2\theta^2\Big\{|z|^2(\varphi_t + 2\lambda\varphi\tilde\alpha_t) + 2\varphi z\Big[\sum_{j,k=1}^n (p^{jk}z_{x_j})_{x_k} + g\Big]\Big\}dxdt \\
&= \int_0^T \int_{\omega_0} \theta^2\Big[\zeta^2|z|^2(\varphi_t + 2\lambda\varphi\tilde\alpha_t) - 2\zeta^2\varphi\sum_{j,k=1}^n p^{jk}z_{x_j}z_{x_k} \\
&\qquad - 2\mu\zeta^2\varphi\sum_{j,k=1}^n p^{jk}zz_{x_j}\psi_{x_k} - 4\lambda\mu\zeta^2\varphi^2\sum_{j,k=1}^n p^{jk}zz_{x_j}\psi_{x_k} \\
&\qquad - 4\zeta\varphi\sum_{j,k=1}^n p^{jk}zz_{x_j}\zeta_{x_k} + 2\zeta^2\varphi zg\Big]dxdt.
\end{aligned} \tag{3.21}
$$

Consequently, we conclude from (3.21) that, for some $\varepsilon > 0$,

$$
\begin{aligned}
2\int_0^T &\int_{\omega_0} \zeta^2\varphi\theta^2 \sum_{j,k=1}^n p^{jk}z_{x_j}z_{x_k}dxdt \\
&\le \varepsilon\int_0^T \int_{\omega_0} \zeta^2\varphi\theta^2|\nabla z|^2 dxdt \\
&\quad + \frac{C}{\varepsilon}\Big(\frac{1}{\lambda^2\mu^2}\int_0^T \int_{\omega_0} \theta^2|g|^2 dxdt + \lambda^2\mu^2\int_0^T \int_{\omega_0} \varphi^3\theta^2|z|^2 dxdt\Big).
\end{aligned} \tag{3.22}
$$

Combining (3.2), (3.20) and (3.22), we have

$$
\begin{aligned}
\int_Q |I_1 + \Phi v|^2 dxdt &+ \lambda\mu^2 \int_Q \varphi\theta^2\big(|\nabla z|^2 + \lambda^2\mu^2\varphi^2|z|^2\big)dxdt \\
&\le C\Big(\int_Q \theta^2|g|^2 dxdt + \lambda^3\mu^4 \int_0^T \int_{\omega_0} \varphi^3\theta^2|z|^2 dxdt\Big).
\end{aligned} \tag{3.23}
$$

Step 3. Finally, let us estimate $\int_Q (\lambda\varphi)^{-1}\theta^2\Big(|z_t|^2 + \Big|\sum_{j,k=1}^{n}(p^{jk}z_{x_j})_{x_k}\Big|^2\Big)dxdt.$

Noting that $v = \theta z$, we see that

$$
\begin{cases}
-\theta g = \theta\Big[-z_t + \displaystyle\sum_{j,k=1}^{n}(p^{jk}z_{x_j})_{x_k}\Big] = I_1 + I_2, \\[4mm]
I_2 = -v_t - 2\displaystyle\sum_{j,k=1}^{n}p^{jk}\ell_{x_j}v_{x_k} + \Psi v + \Phi v,
\end{cases}
\tag{3.24}
$$

where Ψ is given by (3.12), and I_1 and A are given by (3.4). By (3.4), (3.7)–(3.12), (3.13), and (3.24), we have

$$
\int_Q (\lambda\varphi)^{-1}\theta^2\Big(|z_t|^2 + \Big|\sum_{j,k=1}^{n}(p^{jk}z_{x_j})_{x_k}\Big|^2\Big)dxdt
$$

$$
= \int_Q (\lambda\varphi)^{-1}\theta^2\Big(|z_t|^2 + |z_t - g|^2\Big)dxdt
$$

$$
\leq C\int_Q (\lambda\varphi)^{-1}\big(|v_t - \ell_t v|^2 + |\theta g|^2\big)\,dxdt
\tag{3.25}
$$

$$
\leq C\int_Q (\lambda\varphi)^{-1}\big[|I_2 - \Phi v|^2 + (\lambda\mu\varphi)^4|v|^2 + (\lambda\mu\varphi)^2|\nabla v|^2 + |\theta g|^2\big]dxdt
$$

$$
\leq C\Big[\int_Q (\lambda\varphi)^{-1}\big(|I_1 + \Phi v|^2 + |\theta g|^2\big)dxdt + \lambda\mu^2\varphi\int_Q (\lambda^2\mu^2\varphi^2|v|^2 + |\nabla v|^2)dxdt\Big].
$$

Combining (3.23) and (3.25), we complete the proof of Theorem 3.1.

Next, we recall the following known result.

Lemma 3.1 ([24]) *Let $\Gamma_0 \subset \Gamma$ be an arbitrary nonempty open set. Then there exists a function $\tilde\psi \in C^2(\overline\Omega)$ such that*

$$
\tilde\psi > 0 \ in\ \Omega, \quad |\nabla\tilde\psi| > 0 \ in\ \Omega, \quad \tilde\psi|_{\Gamma\setminus\Gamma_0} = 0.
\tag{3.26}
$$

Similarly to Theorem 3.1, we have the following Carleman estimate.

Theorem 3.2 *Let φ and Ψ be given respectively in (3.6) and (3.12), with ψ being replaced by the function $\tilde\psi$ in Lemma 3.1. Then there exists $\tilde\mu_0 > 0$ such that for all $\mu \geq \tilde\mu_0$, one can find two constants $C > 0$ and $\tilde\lambda_1 = \tilde\lambda_1(\mu)$ so that for all $\lambda \geq \tilde\lambda_1$ and $z \in C([0,T]; L^2(\Omega)) \cap L^2(0,T; H_0^1(\Omega))$ solving (3.9), it holds that*

$$\int_Q (\lambda\varphi)^{-1}\theta^2\Big(|z_t|^2 + \Big|\sum_{j,k=1}^n (p^{jk}z_{x_j})_{x_k}\Big|^2\Big)dxdt$$

$$+ \lambda^3\mu^4\int_Q \theta^2\varphi^3|z|^2dxdt + \lambda\mu^2\int_Q \theta^2\varphi|\nabla z|^2dxdt \tag{3.27}$$

$$\leq C\Big(|\theta g|^2_{L^2(Q)} + \lambda\mu\int_0^T\int_{\Gamma_0} \varphi\theta^2\Big|\frac{\partial z}{\partial\nu}\Big|^2 d\Gamma dt\Big).$$

Proof Using Lemma 3.1, and by (3.26) and (3.14)–(3.16), we conclude that there is a $\tilde{\mu}_0 > 0$ such that for all $\mu \geq \tilde{\mu}_0$, one can find two constants $c_1 > 0$ and $\tilde{\lambda}_1 = \tilde{\lambda}_1(\mu)$ so that for all $\lambda \geq \tilde{\lambda}_1$, it holds that

$$\sum_{j,k=1}^n c^{jk}v_{x_j}v_{x_k} \geq c_1\lambda\mu^2\varphi, \quad B \geq c_1\lambda^3\mu^4\varphi^3 \tag{3.28}$$

and

$$\int_\Sigma V \cdot v d\Gamma dt \leq C\lambda\mu\int_0^T\int_{\Gamma_0} \varphi\Big|\frac{\partial v}{\partial\nu}\Big|^2 d\Gamma dt. \tag{3.29}$$

Combining (3.11), (3.25), (3.28) and (3.29), noting that $v = \theta z$, we immediately get the desired result (3.27).

Finally, to study the SUCP for parabolic equations, we shall establish a local Carleman estimate. Without loss of generality, we assume that $0 \in \Omega$ and that $(p^{jk}(t, 0))_{1 \leq j,k \leq n}$ is the identity matrix for all $t \in [0, T]$. Under these assumptions, we choose the weight functions to be the same as that for the elliptic case (See (2.22)–(2.23)).

We have the following result (appeared in [46]).

Theorem 3.3 *Let w be given in (2.23). Then, there exists $\mu_0 > 0$ such that for $\mu = \mu_0$ in (2.23), one can find two constants $\lambda_0 = \lambda_0(\mu) > 0$ and $R_0 > 0$ so that for all $z \in C_0^2\big([(\mathscr{B}(0, R_0)\backslash\{0\}) \times (0, T)] \cap Q\big)$ and $\lambda \geq \lambda_0$, it holds that*

$$\lambda\mu\int_Q \big(w^{1-2\lambda}|\nabla z|^2 + \lambda^2 w^{-1-2\lambda}|z|^2\big)dxdt$$

$$\leq C\int_Q w^{2-2\lambda}\Big|z_t - \sum_{j,k=1}^n (p^{jk}z_{x_j})_{x_k}\Big|^2 dxdt. \tag{3.30}$$

Proof We divide the proof into two steps.

Step 1. For $\sigma = |x| > 0$, let

$$\phi(\sigma) = \frac{\tilde{\varphi}(\sigma)}{\sigma\tilde{\varphi}'(\sigma)}. \tag{3.31}$$

Then, by (2.22), it is easy to see that

$$\phi(\sigma) = e^{\mu\sigma}, \quad \phi'(\sigma) = \mu\phi(\sigma). \tag{3.32}$$

Put

$$\mathscr{T} = \Big(\sum_{j,k=1}^{n} p^{jk}\sigma_{x_j}\sigma_{x_k} \Big)^{-1} = |x|^2 \Big(\sum_{j,k=1}^{n} p^{jk}x_j x_k \Big)^{-1}. \tag{3.33}$$

In (3.3), take $m = n$, $(a^{jk})_{n\times n} = (p^{jk})_{n\times n}$, $\Psi = -\sum_{j,k=1}^{n} (p^{jk}\tilde{\ell}_{x_j})_{x_k}$, $\Phi = 0$ and $\theta(x)$
$= \tilde{\theta}(x)$ (given by (2.23)). Multiplying (3.3) by $(\sigma\phi)^2\mathscr{T}$, we have that

$$(\sigma\phi)^2\mathscr{T}\tilde{\theta}^2 \Big| z_t - \sum_{j,k=1}^{n} (p^{jk}z_{x_j})_{x_k} \Big|^2 + \sum_{k=1}^{n} (\sigma^2\phi^2\mathscr{T}V^k)_{x_k} + (\sigma^2\phi^2\mathscr{T}M)_t$$

$$\geq (\sigma\phi)^2\mathscr{T}|\tilde{I}_1|^2 + 2\sum_{j,k=1}^{n} \tilde{c}^{jk}v_{x_j}v_{x_k} + \sum_{k=1}^{n}(\sigma^2\phi^2\mathscr{T})_k V^k + (\sigma^2\phi^2\mathscr{T})_t M \tag{3.34}$$

$$- 2(\sigma\phi)^2\mathscr{T}\sum_{j,k=1}^{n} p^{jk}\Psi_{x_j}v_{x_k}v + \tilde{B}|v|^2.$$

Here

$$\tilde{I}_1 = \sum_{j,k=1}^{n} (p^{jk}v_{x_j})_{x_k} + \tilde{A}v, \quad \tilde{A} = \sum_{j,k=1}^{n} p^{jk}\tilde{\ell}_{x_j}\tilde{\ell}_{x_k}, \tag{3.35}$$

and

$$\begin{cases}
V^k = \sum_{j=1}^{n} \Big[2p^{jk}v_{x_j}v_t + 2\sum_{j',k'=1}^{n} \Big(2p^{jk'}p^{j'k} - p^{jk}p^{j'k'} \Big)\tilde{\ell}_{x_j}v_{x_{j'}}v_{x_{k'}} \\
\qquad\qquad - 2\Psi p^{jk}v_{x_j}v + 2\tilde{A}p^{jk}\tilde{\ell}_{x_j}v^2 \Big], \\
M = \tilde{A} - \sum_{j,k=1}^{n} p^{jk}v_{x_j}v_{x_k}, \\
\tilde{c}^{jk} = (\sigma\phi)^2\mathscr{T}\sum_{j',k'=1}^{n} \Big[2\big(p^{j'k}\tilde{\ell}_{x_j}\big)_{x_{k'}}p^{jk'} - (p^{jk})_{x_{k'}}p^{j'k'}\tilde{\ell}_{x_{j'}} \Big] - \frac{1}{2}(\sigma\phi)^2\mathscr{T}p_t^{jk}, \\
\tilde{B} = (\sigma\phi)^2\mathscr{T}\Big(2\sum_{j,k=1}^{n} p^{jk}\tilde{\ell}_{x_j}\tilde{A}_{x_k} + \tilde{A}_t \Big).
\end{cases} \tag{3.36}$$

For simplicity, we denote

$$
\begin{cases}
\mathscr{H}_1 = 2 \displaystyle\sum_{j,k,j',k'=1}^{n} (\sigma^2\phi^2\mathscr{T})_{x_k}\left(2p^{jk'}p^{j'k} - p^{jk}p^{j'k'}\right)\tilde{\ell}_{x_j}v_{x_{j'}}v_{x_k'} + 2\sum_{j,k=1}^{n}\tilde{c}^{jk}v_{x_j}v_{x_k}, \\[2mm]
\mathscr{H}_2 = 2\tilde{A}\displaystyle\sum_{j,k=1}^{n} p^{jk}\tilde{\ell}_{x_j}(\sigma^2\phi^2\mathscr{T})_{x_k}v^2 + 2(\sigma\phi)^2\mathscr{T}\sum_{j,k=1}^{n}p^{jk}\tilde{\ell}_{x_j}\tilde{A}_{x_k}v^2, \\[2mm]
\mathscr{H}_3 = -2\displaystyle\sum_{j,k=1}^{n} p^{jk}\Psi v_{x_j}(\sigma^2\phi^2\mathscr{T})_{x_k}v - 2(\sigma\phi)^2\mathscr{T}\sum_{j,k=1}^{n}p^{jk}\Psi_{x_j}v_{x_k}v, \\[2mm]
\mathscr{H}_4 = \left(\sigma^2\phi^2\mathscr{T}\right)_t\tilde{A}v^2 + \sigma^2\phi^2\mathscr{T}\tilde{A}_tv^2, \\[2mm]
\mathscr{H}_5 = 2\displaystyle\sum_{j,k=1}^{n} p^{jk}v_{x_j}v_t(\sigma^2\phi^2\mathscr{T})_{x_k} - \sum_{j,k=1}^{n}p^{jk}v_{x_j}v_{x_k}(\sigma^2\phi^2\mathscr{T})_t.
\end{cases}
\tag{3.37}
$$

Then (3.34) can be rewritten as the following:

$$
(\sigma\phi)^2\mathscr{T}\tilde{\theta}^2\left|z_t - \sum_{j,k=1}^{n}(p^{jk}z_{x_j})_{x_k}\right|^2 + \sum_{k=1}^{n}(\sigma^2\phi^2\mathscr{T}V^k)_{x_k} + (\sigma^2\phi^2\mathscr{T}M)_t
$$
$$
\geq (\sigma\phi)^2\mathscr{T}|\tilde{I}_1|^2 + \sum_{j=1}^{5}\mathscr{H}_j.
\tag{3.38}
$$

Step 2. Let us estimate $\mathscr{H}_j$ ($j = 1, 2, 3, 4, 5$). By (3.33), (3.35) and (3.37), it is clear that $\sigma^2\phi^2\mathscr{T}\tilde{A}$ is independent of time variable t. Hence

$$
\mathscr{H}_4 = \left(\sigma^2\phi^2\mathscr{T}\tilde{A}\right)_t v^2 = 0.
\tag{3.39}
$$

Next, recall (3.33) for $\mathscr{T}$, proceeding exactly the same estimation as that for (2.37), we have that, when $|x| \to 0$,

$$
\mathscr{T}_{x_j} = O(1), \qquad j = 1, \ldots, n,
\tag{3.40}
$$

and

$$
\mathscr{T}_t = |x|^2\left(\sum_{j,k=1}^{n}p^{jk}x_jx_k\right)^{-2}\sum_{j,k=1}^{n}p_t^{jk}x_jx_k = O(1).
\tag{3.41}
$$

Further, recalling (3.35) for $\tilde{I}_1$, it is easy to get that

$$
\tilde{I}_2 = \tilde{\theta}\left[-z_t + \sum_{j,k=1}^{n}(p^{jk}z_{x_j})_{x_k}\right] - \tilde{I}_1
$$
$$
= -v_t - 2\sum_{j,k=1}^{n}p^{jk}\tilde{\ell}_{x_j}v_{x_k} - \sum_{j,k=1}^{n}(p^{jk}\tilde{\ell}_{x_j})_{x_k}v
$$

$$= 2\lambda(\sigma\phi)^{-1} \sum_{j,k=1}^{n} p^{jk} \sigma_{x_j} v_{x_k} - \sum_{j,k=1}^{n} (p^{jk} \tilde{\ell}_{x_j})_{x_k} v - v_t. \tag{3.42}$$

It follows from (2.23) that $\left| \sum_{j,k=1}^{n} (p^{jk} \ell_{x_j})_{x_k} \right| \le C\lambda\mu w^{-2}$. This, together with (3.42), implies that

$$4v_t \sum_{j,k=1}^{n} p^{jk} v_{x_j} \sigma_{x_k} = 2\frac{\sigma\phi}{\lambda} \Big[\tilde{I}_2 + \sum_{j,k=1}^{n} (p^{jk} \ell_{x_j})_{x_k} v + v_t \Big] v_t$$

$$\ge \frac{\sigma\phi}{\lambda} \big(v_t^2 - |\tilde{I}_2|^2 \big) + \Big[\frac{\sigma\phi}{\lambda} \sum_{j,k=1}^{n} (p^{jk} \ell_{x_j})_{x_k} v^2 \Big]_t - C\mu^2 v^2. \tag{3.43}$$

By (3.37), (3.40)–(3.41), (3.43) and (2.46), there exist positive constants C_1 and C_2 such that

$$\mathcal{H}_5 = 4(1 + \mu\sigma)\sigma\phi^2 v_t \sum_{j,k=1}^{n} p^{jk} v_{x_j} \sigma_{x_k} + \sigma^2\phi^2 \sum_{j,k=1}^{n} p^{jk} \big(2v_{x_j} \mathcal{T}_{x_k} v_t - \mathcal{T}_t v_{x_j} v_{x_k} \big)$$

$$\ge \frac{C_1}{\lambda} w^2 v_t^2 - \frac{C_2\mu}{\lambda} w^2 |\tilde{I}_2|^2 - C_2\mu^2 v^2 - C_2\lambda w^2 |\nabla v|^2 \tag{3.44}$$

$$+ \Big[2(1 + \mu\sigma)\sigma\phi^2 \frac{\sigma\phi}{2\lambda} \sum_{j,k=1}^{n} (p^{jk} \ell_{x_j})_{x_k} v^2 \Big]_t,$$

Next, the estimation of $\mathcal{H}_j$ ($j = 1, 2, 3$) are almost the same as that in the proof of Theorem 2.2, and therefore, we do not repeat it here. Integrating (3.38) on Q, by (3.39), (3.42) and (3.44), we conclude that there is a constant $\mu_0 > 0$ such that by choosing $\mu = \mu_0$, one can find $\lambda_1 = \lambda_1(\mu_0) > 0$ so that for all $\lambda \ge \lambda_1$ and $z \in C_0^2(Q\backslash(\{0\} \times (0, T)))$, it holds that

$$\lambda\mu \int_Q w \big(p_0 |\nabla v|^2 + \lambda^2 w^{-2} v^2 \big) dxdt + \int_Q w^2 |\tilde{I}_1|^2 dxdt + \int_Q \frac{w^2}{\lambda} v_t^2 dxdt$$

$$\le C \int_Q w^{2-2\lambda} \Big| z_t - \sum_{j,k=1}^{n} (p^{jk} z_{x_j})_{x_k} \Big|^2 dxdt + C \int_Q \mathcal{H}_6 dxdt, \tag{3.45}$$

where

$$\mathcal{H}_6 = \lambda\mu w \Big| \sum_{j,k=1}^{n} p^{jk} \sigma_{x_j} v_{x_k} \Big|^2 + \lambda\mu\sigma\phi |v| |\tilde{I}_1|. \tag{3.46}$$

By (2.23) and (3.37), we have that

$$
\lambda \mu w \left| \sum_{j,k=1}^{n} p^{jk} \sigma_{x_j} v_{x_k} \right|^2
$$
$$
\leq \frac{C\mu w}{\lambda} \left[w^2 \tilde{\theta}^2 \left| z_t - \sum_{j,k=1}^{n} (p^{jk} z_{x_j})_{x_k} \right|^2 + w^2 |\tilde{I}_1|^2 + \lambda^2 \mu^2 w^{-2} v^2 \right] + \frac{C\mu}{\lambda} w^3 v_t^2.
$$

(3.47)

From the definition of w, we deduce that there is an $R_0 > 0$ such that for all $|x| \leq R_0$,

$$
\frac{C_1}{\lambda} w^2 > \frac{C\mu_0}{\lambda} w^3.
$$

(3.48)

Combining (3.45)–(3.48), we may find a $\lambda_0 > 0$ such that the desired estimate (3.30) holds for $\mu = \mu_0$ and $\lambda \geq \lambda_0$.

3.2   Null Controllability for Semilinear Parabolic Equations

As an important application of the global Carleman estimate (3.10) (see Theorem 3.1), in this section, we study the null controllability problem of the following semilinear parabolic equation:

$$
\begin{cases}
y_t - \sum_{j,k=1}^{n} (p^{jk} y_{x_j})_{x_k} = \chi_\omega u + f(y, \nabla y) & \text{in } Q , \\
y = 0 & \text{on } \Sigma , \\
y(0) = y_0 & \text{in } \Omega
\end{cases}
$$

(3.49)

where y and u are, respectively, the *state* and the *control variables*, and f is a given C^1 function defined on $\mathbb{R}^{1+n}$, $f(0,0) = 0$ and

$$
\lim_{|(s,p)| \to \infty} \frac{\left| \int_0^1 f_s(\tau s, \tau p) d\tau \right|}{\ln^{3/2}(1 + |s| + |p|)} = 0,
$$
$$
\lim_{|(s,p)| \to \infty} \frac{\left| \left(\int_0^1 f_{p_1}(\tau s, \tau p) d\tau, \dots \int_0^1 f_{p_n}(\tau s, \tau p) d\tau \right) \right|}{\ln^{1/2}(1 + |s| + |p|)} = 0,
$$

(3.50)

where $p = (p_1, \dots, p_n) \in \mathbb{R}^n$.

Remark 3.3 Following [11], one can show that the growth conditions on the nonlinearity $f(y, \nabla y)$ in (3.50) are sharp in some sense.

Our aim is to study the null controllability property (*resp.* approximately controllability property) of the parabolic equation (3.49), by which we mean that for any given initial state $y_0 \in L^2(\Omega)$ (*resp.* for any given $\varepsilon > 0$, any initial state $y_0 \in L^2(\Omega)$ and final state $y_1 \in L^2(\Omega)$), find (if possible) a control $u \in L^2(Q)$ such that the corresponding solution satisfies $y(T) = 0$ in Ω (*resp.* $|y(T) - y_1|_{L^2(\Omega)} \leq \varepsilon$).

We have the following controllability result.

Theorem 3.4 ([18]) *For any nonempty open subset ω and $T > 0$, the Eq. (3.49) is null controllable.*

As a consequence of Theorem 3.4, we have the following approximate controllability result for (3.49).

Corollary 3.2 ([18]) *For any nonempty open subset ω and $T > 0$, the Eq. (3.49) is approximately controllable.*

In order to prove the above controllability results, one needs to consider the following adjoint equation of the linearized system of (3.49):

$$\begin{cases} z_t + \sum_{j,k=1}^{n} (p^{jk} z_{x_j})_{x_k} = az + \sum_{k=1}^{n} a_1^k z_{x_k} & \text{in } Q, \\ z = 0 & \text{on } \Sigma, \\ z(T) = z_T & \text{in } \Omega, \end{cases} \tag{3.51}$$

where $z_T \in L^2(\Omega)$, $a \in L^\infty(0, T; L^p(\Omega))$ for some $p \in [n, \infty]$ and $a_1^1, \ldots, a_1^n \in L^\infty(Q)$. Let

$$r_1 = |a|_{L^\infty(0,T;L^p(\Omega))}, \quad r_2 = \sum_{k=1}^{n} |a_1^k|_{L^\infty(Q)}. \tag{3.52}$$

Thanks to the classical duality argument, in order to prove Theorem 3.4, it suffices to establish the following observability estimate.

Theorem 3.5 ([11, 18]) *For any $T > 0$ and any nonempty open subset ω of Ω, there is a constant $C > 0$ such that for all solutions of the Eq. (3.51), it holds that*

$$|z(0)|_{L^2(\Omega)} \leq \exp\left\{C\left[1 + \frac{1}{T} + Tr_1 + r_1^{\frac{1}{3/2-n/p}} + (1 + T)r_2^2\right]\right\}|z|_{L^2((0,T)\times\omega)}, \tag{3.53}$$

where r_1 and r_2 are given in (3.52).

Proof We divide the proof into three steps.

Step 1. By Theorem 3.1 and (3.51), there exists $\mu_0 > 0$ such that for all $\mu \geq \mu_0$, one can find two constants $C > 0$ and $\lambda_1 = \lambda_1(\mu)$ so that for all $\lambda \geq \lambda_1$, it holds that

$$\begin{aligned} &\lambda^3 \mu^4 \int_Q \theta^2 \varphi^3 |z|^2 dx dt + \lambda \mu^2 \int_Q \theta^2 \varphi |\nabla z|^2 dx dt \\ &\leq C\left[\int_Q \theta^2 \left(az + \sum_{k=1}^{n} a_1^k z_{x_k}\right)^2 dx dt + \lambda^3 \mu^4 \int_0^T \int_\omega \varphi^3 \theta^2 |z|^2 dx dt\right]. \end{aligned} \tag{3.54}$$

On the other hand, for any $g \in H^1(\mathbb{R}^n)$, by Hölder's inequality, one has

$$
\begin{aligned}
|g|^2_{H^{n/p}(\mathbb{R}^n)} &= \int_{\mathbb{R}^n} (1 + |\xi|^2)^{n/p} |\hat{g}(\xi)|^{2n/p} |\hat{g}(\xi)|^{2(1-n/p)} d\xi \\
&\leq |g|^{n/p}_{H^1(\mathbb{R}^n)} |g|^{1-n/p}_{L^2(\mathbb{R}^n)}.
\end{aligned}
$$

This yields immediately

$$
|g|^2_{H^{n/p}_0(\Omega)} \leq C |g|^{n/p}_{H^1_0(\Omega)} |g|^{1-n/p}_{L^2(\Omega)}, \quad \forall g \in H^1_0(\Omega), \tag{3.55}
$$

for some constant $C > 0$, independent of g. Hence, for any $h \in L^2(0, T; H^1_0(\Omega))$,

$$
|h|^2_{L^2(0,T;H^{n/p}_0(\Omega))} \leq C |h|^{n/p}_{L^2(0,T;H^1_0(\Omega))} |h|^{1-n/p}_{L^2(Q)}. \tag{3.56}
$$

Now, recalling the definition of r_1 in (3.52), by Hölder's and Sobolev's inequalities, and the inequality (3.56) we have that, for $1/s + 1/p = 1/2$,

$$
\begin{aligned}
|\theta a z|_{L^2(Q)} &\leq r_1 |\theta z|_{L^2(0,T;L^s(\Omega))} \leq r_1 |\theta z|_{L^2(0,T;H^{n/p}_0(\Omega))} \\
&\leq r_1 |\theta z|^{n/p}_{L^2(0,T;H^1_0(\Omega))} |\theta z|^{1-n/p}_{L^2(Q)}.
\end{aligned}
$$

This, together with Young's inequality, implies that, for any $\varepsilon > 0$:

$$
|\theta a z|^2_{L^2(Q)} \leq \varepsilon \lambda |\theta z|^2_{L^2(0,T;H^1_0(\Omega))} + C_\varepsilon r_1^{2p/(p-n)} \lambda^{-n/(p-n)} |\theta z|^2_{L^2(Q)}, \tag{3.57}
$$

where C_ε is a positive constant depending on ε. Combining (3.54) and (3.57), and taking $\varepsilon > 0$ small enough, we get that, for some large constant $C_1 > 0$,

$$
\begin{aligned}
&\lambda^3 \mu^4 \int_Q \theta^2 \varphi^3 |z|^2 dx dt + \lambda \mu^2 \int_Q \theta^2 \varphi |\nabla z|^2 dx dt \\
&\leq C_1 \Big(r_1^{\frac{2p}{p-n}} \lambda^{\frac{-n}{p-n}} \int_Q \theta^2 |z|^2 dx dt + r_2^2 \int_Q \theta^2 |\nabla z|^2 dx dt + \lambda^3 \mu^4 \int_0^T \int_\omega \varphi^3 \theta^2 |z|^2 dx dt \Big).
\end{aligned}
$$

Now, choose $\lambda_1 > 0$ large enough such that

$$
\lambda > \lambda_1 \Big(1 + r_1^{\frac{1}{3/2-n/p}} + r_2^2 \Big) \implies C_1 \Big(r_1^{\frac{2p}{p-n}} + r_2^2 \Big) \leq \frac{\lambda}{2}.
$$

Then,

$$
\begin{aligned}
\int_Q \theta^2 \varphi^3 |z|^2 dx dt &\leq C \int_0^T \int_\omega \varphi^3 \theta^2 |z|^2 dx dt, \\
&\forall \lambda > \lambda_2 \overset{\Delta}{=} \lambda_1 \Big(1 + r_1^{\frac{1}{3/2-n/p}} + r_2^2 \Big).
\end{aligned} \tag{3.58}
$$

Step 2. Recalling (3.6) for the definition of φ and $\tilde{\alpha}$, and $\rho = \psi$ given by Lemma 2.4, we have that

$$\int_{\frac{T}{4}}^{\frac{3T}{4}} \int_{\Omega} e^{2\lambda\tilde{\alpha}} t^{-3} (T-t)^{-3} z^2 dxdt \leq C \int_{0}^{T} \int_{\omega} e^{2\lambda\tilde{\alpha}} \varphi^3 z^2 dxdt, \quad \forall \lambda > \lambda_2. \quad (3.59)$$

Set

$$h(t,x) = e^{-2\lambda\tilde{\alpha}} t^3 (T-t)^3 = \tau^3 e^{\frac{2\lambda\tilde{\alpha}_1}{\tau}} = J(\tau, x) \quad (3.60)$$

where $\tau = t(T-t) \in [0, T^2/4]$ and $\tilde{\alpha}_1(x) = e^{2\mu|\psi|_{C(\overline{\Omega};\,\mathbb{R})}} - e^{\mu\psi(x)}$.

It is easy to see that $\tau \in [3T^2/16, T^2/4]$, provided that $t \in [T/4, 3T/4]$. From $\frac{dJ(\tau,x)}{d\tau} = 0$, we get the critical value $\hat{\tau} = \frac{2}{3}\lambda\tilde{\alpha}_1(x)$ and $J(\hat{\tau}, x) = \left(\frac{2}{3}\lambda\tilde{\alpha}_1(x)\right)^3 e^3$. On the other hand, $J(0, x) = +\infty$, hence $J(\tau, x)$ is decreasing in $(0, \hat{\tau})$. Therefore, when $\hat{\tau} = \frac{2}{3}\lambda\tilde{\alpha}_1(x) \geq 3T^2/16$, i.e., $\lambda \geq \lambda_3 \overset{\Delta}{=} \frac{9T^2}{32}\left(\min_{x \in \overline{\Omega}} \tilde{\alpha}_1(x)\right)^{-1}$, we have that

$$\max_{t \in (\frac{T}{4}, \frac{3T}{4})} h(t,x) \leq \left(\frac{3}{16}T^2\right)^3 e^{C\lambda T^{-2}}, \quad \forall x \in \Omega. \quad (3.61)$$

By (3.60) and (3.61), we find that there are two constants $C > 0$ and $\lambda_3 > 0$, such that for any $\lambda > \lambda_3$, it holds that

$$\min_{t \in (\frac{T}{4}, \frac{3T}{4})} e^{2\lambda\tilde{\alpha}} t^{-3} (T-t)^{-3} \geq C T^{-6} e^{-C\lambda T^{-2}}. \quad (3.62)$$

Step 3. It follows from (3.59) and (3.62) that for all $\lambda > \lambda_4 \overset{\Delta}{=} \max\{\lambda_2, \lambda_3\}$,

$$\int_{\frac{T}{4}}^{\frac{3T}{4}} \int_{\Omega} z^2 dxdt \leq C T^6 e^{C\lambda T^2} \int_{0}^{T} \int_{\omega_1} e^{2\lambda\tilde{\alpha}} \varphi^3 z^2 dxdt. \quad (3.63)$$

Applying the usual energy estimate to the Eq. (3.51), we see that

$$\int_{\Omega} |z(t_1)|^2 dx \leq e^{2(r_1 + r_2^2)(t_2 - t_1)} \int_{\Omega} |z(t_2)|^2 dx, \quad \text{for any } 0 \leq t_1 \leq t_2 \leq T.$$

Therefore,

$$\int_{\Omega} |z(0)|^2 dx \leq e^{\frac{T}{2}(r_1 + r_2^2)} \int_{\Omega} |z(T/4)|^2 dx$$
$$\leq \frac{C}{T} e^{CT(r_1 + r_2^2)} \int_{\frac{T}{4}}^{\frac{3T}{4}} \int_{\Omega} |z(x,t)|^2 dxdt. \quad (3.64)$$

By the definitions of μ and λ, (3.63) and (3.64), we complete the proof of Theorem 3.5.

Remark 3.4 The observability estimate in the form of (3.53) was first given in [18] by virtue of a global Carleman estimate. In [11], it was shown that (3.53) is sharp in some sense.

Remark 3.5 Since solutions to the parabolic equations have an infinite propagation speed, the "waiting" time T can be chosen as small as one likes, and the control domain ω dose not need to satisfy any geometric condition but being open and non-empty. On the other hand, due to the time irreversibility and the strong dissipativity of (3.51), one cannot replace $|z(0)|_{L^2(\Omega)}$ in the inequality (3.53) by $|z_T|_{L^2(\Omega)}$.

Now, we are in a position to prove Theorem 3.4.

Proof of Theorem 3.4. We first prove a null controllability result for the linearized system of (3.49):

$$\begin{cases} y_t - \sum_{j,k=1}^n (p^{jk}y_{x_j})_{x_k} = \chi_\omega u - ay + \sum_{k=1}^n (a_1^k y)_{x_k} & \text{in } Q, \\ y = 0 & \text{on } \Sigma, \\ y(0) = y_0 & \text{in } \Omega. \end{cases} \tag{3.65}$$

Denote by H the completion of $L^2(\Omega)$ with respect to the following norm:

$$|z_T|_H \stackrel{\triangle}{=} \left(\int_0^T \int_\omega z^2 dxdt \right)^{\frac{1}{2}},$$

where z is the solution to (3.51) with the final datum $z_T \in L^2(\Omega)$.

Consider the functional

$$\mathscr{J}(z_T) = \frac{1}{2} \int_0^T \int_\omega |z|^2 dxdt + \int_\Omega z(0)y_0 dx \tag{3.66}$$

over H. Clearly, $\mathscr{J}$ is convex and continuous in H. Furthermore, the observability estimate (given by Theorem 3.5) guarantees the coercivity of $\mathscr{J}$ and the existence of its minimizer.

Denote by $\hat{z}_T = \hat{z}(T) \in H$ the minimizer of $\mathscr{J}$ and by $\hat{z}$ the solution of (3.51) with final datum $\hat{z}_T$. Then for any $\xi_T \in L^2(\Omega)$ and $\delta \in \mathbb{R}$, we have $\mathscr{J}(\hat{z}_T) \leq \mathscr{J}(\hat{z}_T + \delta\xi_T)$. Let ξ be the solution to (3.51) with the final datum ξ_T. Then,

$$\begin{aligned} &\mathscr{J}(\hat{z}_T + \delta\xi_T) - \mathscr{J}(\hat{z}_T) \\ &= \frac{\delta^2}{2} \int_0^T \int_\omega \psi^2 dxdt + \delta \left(\int_0^T \int_\omega \hat{z}\xi dxdt + \int_\Omega \xi(0)y_0 dx \right). \end{aligned} \tag{3.67}$$

Dividing (3.67) by $\delta > 0$ and letting $\delta \to 0$, we obtain that

$$\int_0^T \int_\omega \hat{z}\xi dxdt + \int_\Omega \xi(0)y_0 dx \geq 0.$$

The same calculations with $\delta < 0$ give that

$$\int_0^T \int_\omega \hat{z} \xi \, dx dt + \int_\Omega \xi(0) y_0 dx \le 0.$$

On the other hand, by taking the control $u = \hat{z}$ in (3.65), multiplying (3.65) by ξ and using integration by parts, we get that

$$\int_0^T \int_\omega \hat{z} \xi \, dx dt = \int_\Omega \left[\xi_T y(T) - \xi(0) y_0 \right] dx.$$

It follows from the last three relations that

$$\int_\Omega \xi_T y(T) dx = 0, \quad \forall \xi_T \in L^2(\Omega),$$

which is equivalent to $y(T) = 0$.

Finally, by using the standard fixed point technique, we complete the proof of Theorem 3.4.

Remark 3.6 Based on Theorem 3.2, we can prove a similar boundary observability result as that in Theorem 3.5 (instead of Theorem 3.1). Then, using again the duality argument, we can show the null and approximate controllability result for the parabolic equations with boundary controls (e.g., [21]).

3.3 Inverse Parabolic Problems

As an application of Theorem 3.2, in this section, we consider an inverse source problem for the following parabolic equation:

$$\begin{cases} y_t - \sum_{j,k=1}^n (p^{jk} y_{x_j})_{x_k} = q_1 \cdot \nabla y + q_2 y + Rf & \text{in } Q, \\ y = 0 & \text{on } \Sigma. \end{cases} \tag{3.68}$$

Here, $q_1 \in W^{1,\infty}(0, T; L^\infty(\Omega; \mathbb{R}^n))$, $q_2 \in W^{1,\infty}(0, T; L^\infty(\Omega))$ and $R \in W^{1,\infty}(0, T; L^\infty(\Omega))$ are known functions, while $f \in L^2(\Omega)$ is unknown.

We are interested in the following problem:

Problem (IP): Let $\Gamma_0 \subset \Gamma$ be an arbitrarily fixed sub-boundary and let $t_0 \in (0, T)$ be any fixed time. Determine $f(\cdot)$ by $y(t_0, \cdot)$ in Ω and $\sum_{j,k=1}^n p^{jk} y_{x_j} \nu^k$ on $(0, T) \times \Gamma_0$.

Based on Theorem 3.2, we can prove the following result:

Theorem 3.6 *Let* $y, y_t \in H^1(0, T; H^2(\Omega))$. *Assume that*

$$R(t_0, x) \neq 0, \qquad a.e.\ x \in \Omega. \tag{3.69}$$

Then there exists a constant $C > 0$ such that,

$$|f|_{L^2(\Omega)} \leq C \Big(|y(t_0, \cdot)|_{H^2(\Omega)} + \Big| \sum_{j,k=1}^{n} p^{jk} y_{t,x_j} v^k \Big|_{L^2((0,T)\times\Gamma_0)} \Big), \quad \forall f \in L^2(\Omega). \tag{3.70}$$

Proof It is sufficient to prove that

$$|f|_{L^2(\Omega)} \leq C \Big(|y(t_0, \cdot)|_{H^2(\Omega)} + \Big| \sum_{j,k=1}^{n} p^{jk} y_{t,x_j} v^k \Big|_{L^2((t_0-t_1,t_0+t_1)\times\Gamma_0)} \Big) \tag{3.71}$$

with $0 < t_0 - t_1 < t_0 + t_1 < T$. Therefore, by a simple change of variable, it is sufficient to prove (3.70) for $t_0 = \frac{T}{2}$.

It follows from (3.68) that for a.e. $x \in \Omega$,

$$R(T/2, x) f(x) = y_t(T/2, x) - \sum_{j,k=1}^{n} (p^{jk} y_{x_j})_{x_k}(T/2, x) \tag{3.72}$$
$$- q_1(T/2, x) \cdot \nabla y(T/2, x) - q_2 y(T/2, x).$$

In terms of (3.72), we have to estimate $y_t(T/2, x)$. The estimation is based on Theorem 3.2. We first deal with $\int_{\Omega} \theta^2\big(\frac{T}{2}, x\big) \big| y_t\big(\frac{T}{2}, x\big) \big|^2 dx$. Since $\lim_{t\to 0} \theta(x, t) = 0$ for $x \in \Omega$, we have

$$\int_{\Omega} \theta^2\Big(\frac{T}{2}, x\Big) \Big| y_t\Big(\frac{T}{2}, x\Big) \Big|^2 dx = \int_0^{\frac{T}{2}} \frac{\partial}{\partial t}\Big(\int_{\Omega} \theta^2 |y_t|^2 dx \Big) dt$$
$$= 2 \int_{\Omega} \int_0^{\frac{T}{2}} \theta^2 \big(y_t y_{tt} + \ell_t |y_t|^2 \big) dx dt \tag{3.73}$$
$$\leq 2 \int_{\Omega} \int_0^T \theta^2 \big(|y_t||y_{tt}| + C\lambda\varphi^2 |y_t|^2 \big) dx dt$$
$$\leq C \int_{\Omega} \int_0^T \theta^2 \Big(\lambda^2 \varphi^2 |y_t|^2 + \frac{1}{\lambda^2 \varphi} |y_{tt}|^2 \Big) dx dt.$$

Next, for $\mu > 1$, by Theorem 3.2 (with z replaced by y_t in (3.27)), we get that

$$\int_Q \frac{1}{\lambda\varphi} \theta^2 |y_{tt}|^2 dx dt + \lambda^3 \int_Q \theta^2 \varphi^3 |y_t|^2 dx dt + \lambda \int_Q \theta^2 \varphi |\nabla y_t|^2 dx dt$$
$$\leq C \Big(\int_Q \theta^2 |R_t f|^2 dx dt + \lambda\mu \int_0^T \int_{\Gamma_0} \varphi \theta^2 \Big| \sum_{j,k=1}^{n} p^{jk} y_{t,x_j} v^k \Big|^2 d\Gamma dt \Big) \tag{3.74}$$

for all $\lambda > 0$ large enough.

It follows from (3.73) and (3.74) that

$$
\begin{aligned}
&\int_{\Omega} \theta^2\left(\frac{T}{2}, x\right)\left|y_t\left(\frac{T}{2}, x\right)\right|^2 dx \\
&\leq \frac{C}{\lambda} \int_Q \theta^2 |R_t f|^2 dx dt + C e^{C\lambda} \int_0^T \int_{\Gamma_0} \left|\sum_{j,k=1}^n p^{jk} y_{t,x_j} \nu^k\right|^2 d\Gamma dt.
\end{aligned}
\tag{3.75}
$$

By (3.69) and (3.72), we have

$$
\begin{aligned}
\int_{\Omega} \theta^2\left(\frac{T}{2}, x\right)|f(x)|^2 dx &\leq \frac{C}{\lambda} \int_Q \theta^2 |f(x)|^2 dx dt + C e^{c\lambda} \left|y\left(\frac{T}{2}, x\right)\right|_{H^2(\Omega)}^2 \\
&\quad + C e^{C\lambda} \int_0^T \int_{\Gamma_0} \left|\sum_{j,k=1}^n p^{jk} y_{t,x_j} \nu^k\right|^2 d\Gamma dt.
\end{aligned}
\tag{3.76}
$$

Recalling the definition of $\tilde{\alpha}$ in (3.6), we find that

$$
\tilde{\alpha}(t, x) \leq \tilde{\alpha}(T/2, x), \quad \forall (x, t) \in Q.
$$

Thus,

$$
\int_Q \theta^2 |f(x)|^2 dx dt \leq T \int_{\Omega} \theta^2\left(\frac{T}{2}, x\right)|f(x)|^2 dx.
\tag{3.77}
$$

Finally, by (3.76) and (3.77), we get that

$$
\begin{aligned}
&\left(1 - \frac{C}{\lambda}\right) \int_{\Omega} \theta^2\left(x, \frac{T}{2}\right)|f(x)|^2 dx \\
&\leq C e^{c\lambda} \left|y\left(x, \frac{T}{2}\right)\right|_{H^2(\Omega)}^2 + C e^{C\lambda} \int_0^T \int_{\gamma_0} \left|\sum_{j,k=1}^n p^{jk} y_{t,x_j} \nu^k\right|^2 d\Gamma dt.
\end{aligned}
\tag{3.78}
$$

Taking $\lambda > 0$ sufficiently large, we complete the proof of Theorem 3.6.

3.4 Strong Unique Continuation Property of Parabolic Equations

This section is devoted to showing SUCP for solutions to the following equation:

$$
y_t - \sum_{j,k=1}^n (p^{jk} y_{x_j})_{x_k} = a \cdot \nabla y + by \quad \text{in } Q.
\tag{3.79}
$$

Here $a \in L_{loc}^{\infty}(Q; \mathbb{R}^n)$ and $b \in L_{loc}^{\infty}(Q)$.

Definition 3.1 A solution y of (3.79) is said to satisfy SUCP if $y = 0$ in Q provided that, there exist some $x_0 \in \Omega$ and $R > 0$ so that for every $N \in \mathbb{N}$, there is a $C_N > 0$ such that $\int_{(0,T) \times \mathscr{B}(x_0,r)} |y(t,x)|^2 dx dt \leq C_N r^{2N}, \ \forall \, r \in (0, R]$.

The following result holds:

Theorem 3.7 ([7, 46]) *Let $y \in H^1(0, T; H^2_{loc}(\Omega))$ be a solution of (3.79). Then y satisfies SUCP.*

Proof Without loss of generality, we assume that $x_0 = 0$. For $\delta \in (0, T/2)$ and $r > 0$, put

$$\mathscr{Q}(\delta, r) = (\delta, T - \delta) \times \mathscr{B}(0, r). \tag{3.80}$$

Let $\delta_0 > 0$ and $r_0 \in \big(0, \min\{R_0, R\}\big]$ be such that $\mathscr{Q}(\delta_0, r_0) \subset Q$. Let us first introduce two cut-off functions:

$$\rho_1 \in C_0^\infty(\mathscr{Q}(3\delta_0/4, 3r_0/4); [0, 1]), \quad \rho_1 \equiv 1 \text{ in } \mathscr{Q}(\delta_0/2, r_0/2)$$

and

$$\rho_2 \in C^\infty(\mathbb{R}^n; [0, 1]), \quad \rho_2 \equiv 1 \text{ in } \mathbb{R}^n \backslash \mathscr{B}(0, 1), \quad \rho_2 \equiv 0 \text{ in } \mathscr{B}(0, 1/2).$$

For $\varepsilon \in (0, \min\{1, \delta_0/2\})$, put

$$\eta_\varepsilon(t, x) = \rho_2\Big(\frac{x}{\varepsilon}\Big)\rho_1(t, x), \quad y_\varepsilon(t, x) = \eta_\varepsilon(t, x)y(t, x). \tag{3.81}$$

By Theorem 3.3, we obtain that

$$
\lambda^3 \int_{\mathscr{Q}(\delta_0, r_0)} w^{-1-2\lambda} y_\varepsilon^2 dx dt
$$
$$
\leq C \int_{\mathscr{Q}(\delta_0, r_0)} w^{2-2\lambda} \Big| y_{\varepsilon,t} - \sum_{j,k=1}^n (p^{jk} y_{\varepsilon,x_j})_{x_k} \Big|^2 dx dt \tag{3.82}
$$
$$
\leq C \int_{\mathscr{Q}(\delta_0, r_0)} w^{2-2\lambda} \Big| \eta_\varepsilon b y + \sum_{j,k=1}^n [2p^{jk} y_{x_j} \eta_{\varepsilon,x_k} + y(p^{jk}\eta_{\varepsilon,x_j})_{x_k}] - \eta_{\varepsilon,t} y - a \nabla \eta_\varepsilon y \Big|^2 dx dt,
$$

where $w(\cdot)$ is given by (2.23).

It follows from (3.81) that

$$\eta_{\varepsilon,x_k}(t, x) = \frac{1}{\varepsilon}\rho_{2,x_k}\Big(\frac{x}{\varepsilon}\Big)\rho_1(t, x) + \rho_2\Big(\frac{x}{\varepsilon}\Big)\rho_{1,x_k}(t, x),$$

and that

$$\eta_{\varepsilon,x_j x_k}(t,x) = \frac{1}{\varepsilon^2}\rho_{2,x_j x_k}\left(\frac{x}{\varepsilon}\right)\rho_1(t,x) + \rho_2\left(\frac{x}{\varepsilon}\right)\rho_{1,x_j x_k}(t,x)$$
$$+ \frac{1}{\varepsilon}\rho_{2,x_j}\left(\frac{x}{\varepsilon}\right)\rho_{1,x_k}(t,x) + \frac{1}{\varepsilon}\rho_{2,x_k}\left(\frac{x}{\varepsilon}\right)\rho_{1,x_j}(t,x).$$

Hence,

$$\int_{\mathscr{Q}(\delta_0,r_0)} w^{2-2\lambda}\left|\sum_{j,k=1}^{n} h^{jk} y_{x_j}\eta_{\varepsilon,x_k}\right|^2 dxdt$$
$$\leq 2\int_{\mathscr{Q}(\delta_0,r_0)} w^{2-2\lambda}\left|\frac{1}{\varepsilon}\sum_{j,k=1}^{n} h^{jk} y_{x_j}\rho_{2,x_k}\rho_1\right|^2 dxdt$$
$$+ 2\int_{\mathscr{Q}(\delta_0,r_0)} w^{2-2\lambda}\left|\sum_{j,k=1}^{n} h^{jk} y_{x_j}\rho_2\rho_{1,x_k}\right|^2 dxdt.$$

By (2.115), we obtain that

$$\int_{\mathscr{Q}(\delta_0,r_0)} w^{2-2\lambda}\left|\frac{1}{\varepsilon}\sum_{j,k=1}^{n} h^{jk} y_{x_j}\rho_{2,x_k}\rho_1\right|^2 dxdt$$
$$\leq C\int_{\mathscr{Q}(\delta_0,\varepsilon)\backslash\mathscr{Q}(\delta_0,\frac{\varepsilon}{2})} \frac{1}{\varepsilon^2} w^{2-2\lambda}|\nabla y|^2 dxdt$$
$$\leq \frac{C}{\varepsilon^2}\left(\frac{\varepsilon}{2C_1}\right)^{2-2\lambda}\int_{\mathscr{Q}(\delta_0,\varepsilon)\backslash\mathscr{Q}(\delta_0,\frac{\varepsilon}{2})} |\nabla y|^2 dxdt \tag{3.83}$$
$$\leq \frac{C}{\varepsilon^2}\left(\frac{\varepsilon}{2C_1}\right)^{2-2\lambda}\left(|b|_{L^\infty(\mathscr{Q}(\delta_0,2\varepsilon))} + \frac{C}{\varepsilon^2}\right)\int_{\mathscr{Q}(\delta_0,2\varepsilon)\backslash\mathscr{Q}(\delta_0,\frac{\varepsilon}{4})} |y|^2 dxdt$$
$$\leq CC_N\left(|b|_{L^\infty(\mathscr{Q}(\delta_0,r_0))} + 1\right)C_1^{2\lambda-2}2^{2\lambda-2+2N}\varepsilon^{2N-2\lambda-2},$$

where we have used the fact that $\displaystyle\int_{\mathscr{Q}(\delta_0,r)} |y|^2 dx \leq C_N r^{2N}$ for all $N \geq 0$ and $r \in (0,R]$.

Let us choose N large enough such that the right hand side of (3.83) tends to 0 as $\varepsilon \to 0$. Furthermore, proceeding exactly the same analysis as that in (3.83), we can show that all the terms on the right hand side of (3.82), in which ρ_2 is differentiated, tend to 0 as $\varepsilon \to 0$. Consequently, letting $\varepsilon \to 0$, we obtain that

$$\lambda^3\int_{\mathscr{Q}(\delta_0,r_0)} w^{-1-2\lambda}|\rho_1 y|^2 dxdt$$
$$\leq C\int_{\mathscr{Q}(\delta_0,r_0)} w^{2-2\lambda}\left|\rho_1 by + y\sum_{j,k=1}^{n}(h^{jk}\rho_{1,x_j})_{x_k} + 2\sum_{j,k=1}^{n} h^{jk} y_{x_j}\rho_{1,x_k}\right|^2 dxdt.$$
$$\tag{3.84}$$

By (2.23),

$$w(x)^{2-2\lambda} = w(x)^3 w(x)^{-1-2\lambda} \le (C_1\delta_0)^3 w(x)^{-1-2\lambda}, \qquad \forall \, (t, x) \in \mathcal{Q}(\delta_0, r_0).$$

Let λ be large enough such that

$$Cr_0^3 |b|_{L^\infty(\mathcal{Q}(\delta_0, r_0))}^2 \le \frac{\lambda^3}{2}.$$

It follows from (3.84), (2.23) and the monotonicity of $\tilde{\varphi}$ in (2.22) that

$$\frac{\lambda^3}{2} \int_{\mathcal{Q}(\delta_0, r_0)} w^{-1-2\lambda} |\rho_1 y|^2 dx dt$$

$$\le C \int_{\mathcal{Q}(\delta_0, r_0)} w^{2-2\lambda} \left| y \sum_{j,k=1}^{n} (h^{jk}\rho_{1,x_j})_{x_k} + 2 \sum_{j,k=1}^{n} h^{jk} y_{x_j} \rho_{1,x_k} \right|^2 dx dt$$

$$\le C\tilde{\varphi}^{2-2\lambda}\left(\frac{\delta_0}{2}\right) \int_{\mathcal{Q}(\delta_0, 3r_0/4)\backslash\mathcal{Q}(\delta_0, r_0/2)} (|y|^2 + |\nabla y|^2) dx dt$$

$$\le C\tilde{\varphi}^{-1-2\lambda}\left(\frac{\delta_0}{2}\right) \int_{\mathcal{Q}(\delta_0, r_0)} |y|^2 dx dt$$

$$\le CC_0^2\tilde{\varphi}^{-1-2\lambda}\left(\frac{\delta_0}{2}\right),$$

where we use Lemma 2.6 once again. Consequently,

$$\frac{\lambda^3}{2}\tilde{\varphi}\left(\frac{\delta_0}{2}\right)^{1+2\lambda} \int_{\mathcal{Q}(\delta_0, r_0/2)} w(x)^{-1-2\lambda} |y|^2 dx dt \le CC_0^2. \tag{3.85}$$

Since $w(x) = \tilde{\varphi}(|x|)$ and $\tilde{\varphi}$ is an increasing function, we see that $w(x)\tilde{\varphi}(\frac{\delta_0}{2})^{-1} < 1$ on $\mathcal{Q}(\delta_0, r_0/2)$. By letting $\lambda \to \infty$, it follows from (3.85) that $y = 0$ on $\mathcal{Q}(\delta_0, r_0/2)$. Then, similar to the proof of Lemma 2.5, a chain of balls argument shows that $y = 0$ in Q.

3.5   Three Cylinders Inequality of Parabolic Equations

Without loss of generality, we assume that $0 \in \Omega$. In what follows, for $\delta > 0$ and $r > 0$, put

$$Q_{\delta,r} = (-\delta, \delta) \times \mathcal{B}(0, r). \tag{3.86}$$

By Theorem 3.3, we can obtain the following interpolation inequality, which is a modification of the corresponding result in [46].

Theorem 3.8 *Let $r_0 \in (0, R_0]$. Assume that $z \in H^1(-\delta_0, \delta_0; H^2(\mathcal{B}(0, r_0)))$ satisfies*

$$z_t - \sum_{j,k=1}^n (p^{jk} z_{x_j})_{x_k} = \hat{a} \cdot \nabla z + \hat{b} z \quad \text{in } Q_{\delta_0, r_0}, \tag{3.87}$$

where $\hat{a} \in L^\infty(Q_{\delta_0, r_0}; \mathbb{R}^n)$ and $\hat{b} \in L^\infty(Q_{\delta_0, r_0})$. Then there exists a constant $C > 1$ such that for all $0 < r_2 < r_1 < r_0$ and $0 < \delta_1 < \delta_0$, it holds that

$$|z|_{L^2(Q_{\delta_1, r_1})} \le C |z|_{L^2(Q_{\delta_0, r_2})}^{\varepsilon_0} |z|_{L^2(Q_{\delta_0, r_0})}^{1-\varepsilon_0}, \tag{3.88}$$

where

$$\varepsilon_0 = \frac{\ln r_1 - \ln r_2}{\ln r_1 - \ln(r_2/2)}.$$

Proof We borrow some ideas from [46]. The proof is divided into four steps.

Step 1. Let $t_1 \in (0, (\delta_0 - \delta_1)/2)$. Set $T_1 = \delta_0 - t_1/2$ and $T_2 = \delta_0 - t_1$. Choose a cut-off function $\sigma(\cdot) \in C_0^2([-\delta_0, \delta_0])$ be such that

$$\sigma(t) = \begin{cases} 0, & \text{if } t \in [-\delta_0, -T_1] \cup [T_1, \delta_0], \\ 1, & \text{if } t \in [-T_2, +T_2], \\ \exp\left(-\dfrac{\delta_0^3(T_2 - t)^4}{(T_1 - t)^3(T_1 - T_2)^4}\right), & \text{if } t \in (T_2, T_1), \\ \exp\left(-\dfrac{\delta_0^3(T_2 + t)^4}{(T_1 + t)^3(T_1 - T_2)^4}\right), & \text{if } t \in (-T_1, -T_2). \end{cases} \tag{3.89}$$

Let $\alpha_0 \in (0, r_0/6)$ and $f \in C_0^2([0, r_0])$ be such that

$$f(r) = \begin{cases} 0, & \text{if } r \in [0, \alpha_0] \cup [3r_0/4, r_0], \\ 1, & \text{if } r \in [3\alpha_0/2, r_0/2], \end{cases} \tag{3.90}$$

and that

$$\begin{cases} |f'| \le C_f/\alpha_0, & |f''| \le C_f/\alpha_0^2 \quad \text{in } [\alpha_0, 3\alpha_0/2], \\ |f'| \le C_f/r_0, & |f''| \le C_f/r_0^2 \quad \text{in } [r_0/2, 3r_0/4], \end{cases} \tag{3.91}$$

where C_f is a constant.

Put

$$\zeta(t, x) = \sigma(t) f(|x|), \quad \text{for all } (t, x) \in Q_{\delta_0, r_0}. \tag{3.92}$$

Let $\{z_n\}_{n=1}^\infty$ be a sequence in $C_0^\infty(Q_{\delta_0, r_0})$, which converges to z in $H^1(-\delta_0, \delta_0; H^2(\mathcal{B}(0, r_0)))$. By applying the inequality (3.30) (by simply using change of variable with respect to the time variable) to $z_n \zeta$ and passing to the limit, we obtain that

$$\int_{Q_{\delta_0, r_0}} \left(\lambda w^{1-2\lambda} |\nabla(z\zeta)|^2 + \lambda^3 w^{-1-2\lambda} z^2 \zeta^2\right) dx dt$$
$$\le C \int_{Q_{\delta_0, r_0}} w^{2-2\lambda} \left| (z\zeta)_t - \sum_{j,k=1}^n [p^{jk}(z\zeta)_{x_j}]_{x_k} \right|^2 dx dt, \quad \forall \lambda \ge \lambda_0. \tag{3.93}$$

Step 2. To estimate the terms in both side of (3.93), let us divide Q_{δ_0, r_0} into several parts according to the weight function. Put

$$
\mathcal{K}_1 = \left\{ (t, x) \in \mathbb{R}^{1+n} : \frac{3}{2}\alpha_0 \le |x| \le \frac{r_0}{2}, \ t \in [-T_1, -T_2] \cup [T_2, T_1] \right\},
$$

$$
\mathcal{K}_2 = \left\{ (x, t) \in \mathbb{R}^{n+1} : \frac{3}{2}\alpha_0 \le |x| \le \frac{r_0}{2}, \ t \in [-T_2, T_2] \right\},
$$

$$
\mathcal{K}_3 = \left\{ (t, x) \in \mathbb{R}^{1+n} : \frac{r_0}{2} \le |x| \le \frac{3r_0}{4}, \ t \in [-T_1, T_1] \right\},
$$

$$
\mathcal{K}_4 = \left\{ (t, x) \in \mathbb{R}^{1+n} : \alpha_0 \le |x| \le \frac{3\alpha_0}{2}, \ t \in [-T_1, T_1] \right\},
$$

$$
\mathcal{K}_5 = Q_{\delta_0, r_0} \backslash \bigcup_{j=1}^{4} \mathcal{K}_j.
$$

Clearly, it holds that $Q_{\delta_0, r_0} = \bigcup_{j=1}^{5} \mathcal{K}_j$ and $\mathcal{K}_j \cap \mathcal{K}_k = \emptyset$ for $j \ne k$, $j, k = 1, \dots, 5$.

Write

$$
M = |\hat{a}|_{L^\infty(Q_{\delta_0, r_0}; \mathbb{R}^n)} + |\hat{b}|_{L^\infty(Q_{\delta_0, r_0})}.
$$

From (3.89)–(3.93) and (3.87), we obtain that for every $\lambda \ge \lambda_0$,

$$
\int_{\mathcal{K}_1 \cup \mathcal{K}_2} \left(\lambda w^{1-2\lambda} |\nabla(z\zeta)|^2 + \lambda^3 w^{-1-2\lambda} z^2 \zeta^2 \right) dx dt
$$
$$
\le \mathcal{J}_1 + CM^2 \int_{\mathcal{K}_1 \cup \mathcal{K}_2} w^{2-2\lambda} (z^2 + |\nabla z|^2) dx dt + C \int_{\mathcal{K}_1} w^{2-2\lambda} |\zeta_t z|^2 dx dt,
$$
(3.94)

where

$$
\mathcal{J}_1 = CM^2 \int_{\mathcal{K}_3 \cup \mathcal{K}_4} w^{2-2\lambda} (|\nabla z|^2 + z^2) dx dt. \tag{3.95}
$$

By (3.94), there exists $\lambda_1 \ge \lambda_0$ such that for all $\lambda \ge \lambda_1$,

$$
\int_{\mathcal{K}_2} \left(\lambda w^{1-2\lambda} |\nabla z|^2 + \lambda^3 w^{-1-2\lambda} z^2 \right) dx dt \le \mathcal{J}_1 + C \int_{\mathcal{K}_1} \mathcal{E}(t, x; \lambda) z^2 dx dt, \tag{3.96}
$$

where

$$
\mathcal{E}(t, x; \lambda) = \sigma(t)^2 w^{-2-2\lambda} \left[\left(\frac{\sigma'(t)}{\sigma(t)} \right)^2 w^4(x) - \lambda^3 \right].
$$

Step 3. Let us estimate $\int_{\mathcal{K}_1} \mathcal{E}(t, x; \lambda) z^2 dx dt$. For simplicity, we only give the related estimation for $t \in [-T_1, -T_2]$ (Similarly, the estimation for $t \in [T_2, T_1]$ can be obtained). Recalling (3.89) for the definition of $\sigma(t)$, by some elementary calculation, we find that there is a constant $C_1 > 0$ such that for all $\lambda \ge \lambda_1$,

$$\mathscr{E}(t, x \, ; \lambda) \leq \lambda^3 w^{-2-2\lambda} \sigma^2(t) \left[-\frac{1}{2} + \frac{C_1 \sigma_0^6}{(T_1 + t)^8} \frac{w^4}{\lambda^3} \right] \quad \text{in } \mathscr{K}_1. \tag{3.97}$$

Let

$$\mathscr{K}_{1,\lambda} \triangleq \left\{ (t, x) \in \mathscr{K}_1 : -\frac{1}{2} + \frac{C_1 \sigma_0^6}{(T_1 + t)^8} \frac{w^4}{\lambda^3} \geq 0 \right\}, \tag{3.98}$$

where C_1 is the same constant appeared in the right hand side of (3.97).

On one hand, it follows from (3.98) that

$$\lambda^3 \leq \frac{2C_1 \sigma_0^6 w^4}{(T_1 + t)^8}, \quad \frac{T_1 + t}{\sigma_0} \leq \left(\frac{2C_1 w^4}{\lambda^3 \sigma_0^2} \right)^{1/8} \quad \text{in } \mathscr{K}_{1,\lambda}. \tag{3.99}$$

On the other hand, if $\lambda^3 \geq \frac{2^{17} C_1 \sigma_0^6 w^4}{t_1^8}$, we have that

$$|T_1 + t| \geq \frac{T_1 - T_2}{2} = t_1, \quad |T_2 + t| \geq \frac{T_1 - T_2}{2} = t_1 \quad \text{in } \mathscr{K}_{1,\lambda}. \tag{3.100}$$

Finally, by (3.89), (3.99) and (3.100), we conclude that there exists $\lambda_2 \geq \lambda_1$ such that for all $\lambda \geq \lambda_2$,

$$\int_{\mathscr{K}_1} \mathscr{E}(t, x \, ; \lambda) z^2 dx dt \leq \frac{C \sigma_0^6}{t_1^8} \int_{\mathscr{K}_1} \sigma(t)^2 w^{2-2\lambda} z^2 dx dt$$
$$\leq C \sigma_0^{2-2\lambda} \int_{\mathscr{K}_1} z^2 dx dt. \tag{3.101}$$

Step 4. Recalling the definitions of $\mathscr{K}_3$ and $\mathscr{K}_4$, by (3.95), we have that

$$\mathscr{J}_1 \leq C \left(\frac{r_0}{2} \right)^{2-2\lambda} \int_{\mathscr{K}_3} (|\nabla z|^2 + z^2) dx dt + C \alpha_0^{2-2\lambda} \int_{\mathscr{K}_4} (|\nabla z|^2 + z^2) dx dt. \tag{3.102}$$

Let $r \in (3\alpha_0/2, r_0/4)$ and denote by $\mathscr{K}_2^{(r)}$ the set $\{(t, x) \in \mathscr{K}_2 : |x| \leq r\}$. Let $r_1 = r_0/2$. By (3.96), (3.101) and (3.102), we obtain that for all $\lambda \geq \lambda_2$,

$$\int_{\mathscr{K}_2^{(r)}} z^2 dx dt \leq \left(\frac{r_1}{2} \right)^{2\lambda+1} \int_{\mathscr{K}_2} z^2 w^{-1-2\lambda} dx dt$$
$$\leq C \left(\frac{r_1}{2} \right)^{2\lambda+1} \left[\alpha_0^{2-2\lambda} \int_{\mathscr{K}_4} (|\nabla z|^2 + z^2) dx dt + \sigma_0^{2-2\lambda} \int_{\mathscr{K}_1} z^2 dx dt \right. \tag{3.103}$$
$$\left. + r_1^{2-2\lambda} \int_{\mathscr{K}_3} (|\nabla z|^2 + z^2) dx dt \right].$$

Let $\zeta_1 \in C_0^\infty(Q_{\delta_0, 2\alpha_0} \setminus Q_{\delta_0, \alpha_0/2})$ be such that $\zeta_1 = 1$ in $\mathscr{K}_4$. Multiplying the Eq. (3.87) by $\zeta_1 z$, integrating it in $Q_{\delta_0, 2\alpha_0} \setminus Q_{\delta_0, \alpha_0/2}$, and using integration by part, we find that

$$\int_{\mathcal{K}_4} |\nabla z|^2 dx dt \le C\alpha_0^{-2} \int_{Q_{\delta_0, 2\alpha_0} \setminus Q_{\delta_0, \alpha_0/2}} z^2 dx dt. \tag{3.104}$$

Similarly,

$$\int_{\mathcal{K}_3} |\nabla z|^2 dx dt \le C r_0^{-2} \int_{Q_{\delta_0, r_0} \setminus Q_{\delta_0, r_0/3}} z^2 dx dt. \tag{3.105}$$

Put $\eta = |z|_{L^2(Q_{\delta_0, r_2})}$ and $\tilde{\eta} = |z|_{L^2(Q_{\delta_0, r_1})}$. Take $\alpha_0 = r_2/2$. Adding both sides of (3.103) by $\int_{Q_{\delta_1, r_1}} z^2 dx dt$, from (3.104) and (3.105), we obtain that

$$|z|_{L^2(Q_{\delta_1, r_1})}^2 \le C\left[\left(\frac{r_1}{r_2}\right)^{2\lambda-2}\eta^2 + 2^{2-2\lambda}\tilde{\eta}^2\right]. \tag{3.106}$$

Set

$$\lambda_3 = \frac{\ln \tilde{\eta} - \ln \eta}{\ln r_1 - \ln(r_2/2)} + 1. \tag{3.107}$$

If $\lambda_3 \ge \lambda_2$, then, by choosing $\lambda = \lambda_3$ in (3.106), we get that

$$|z|_{L^2(Q_{\delta_1, r_1})} \le C\eta^{\varepsilon_0}\tilde{\eta}^{1-\varepsilon_0}, \tag{3.108}$$

where

$$\varepsilon_0 = \frac{\ln r_1 - \ln r_2}{\ln r_1 - \ln(r_2/2)}.$$

If $\lambda_3 < \lambda_2$ then, by (3.107), we have that

$$|z|_{L^2(Q_{\delta_1, r_1})} \le \tilde{\eta} \le \left(\frac{2r_1}{r_2}\right)^{\lambda-1}\eta.$$

This, together with (3.106), implies (3.108).

3.6 Further Comments

To end this chapter, some comments are in order.

- In this book, we assume that the coefficients of the principal part of the operator is C^1. When these coefficients are discontinuous, the Carleman estimate was established only for some special cases, that is, the discontinuities occurred as jumps at an interface with homogeneous or non-homogeneous transmission conditions. For these cases, one has to modify the method in this book to derive Carleman estimates. The details are too long to be presented here. We refer the interested readers to [8, 9, 29, 30].

- In this book, we assume that the operator is uniformly parabolic. Based on the inequality (3.3), one can also derive some Carleman estimate for degenerate parabolic operators. We omit it here due to the limitation of space. Interested readers are referred to [6, 20] and the references therein.
- The controllability problem of parabolic equations has been studied comprehensively for many years (e.g. [10, 11, 17–19, 21, 42, 52]). In [25], a global Carleman estimate was first employed to study the observability problem for parabolic equations. In this chapter, for simplicity, we only consider the control of a single equation. The method can also used to handle some coupled parabolic equations (e.g. [11, 23]).
- Combining Hardy's inequality, we can apply the Carleman estimate (in Theorem 3.1) to establish the null/approximate controllability for parabolic equations with some singular potential (e.g., [3, 12, 45]).
- An early work on inverse problems by means of Carleman estimate is [5]. As for further works for the parabolic case, we mention [27] on the uniqueness and [24] on the Lipschitz stability, and we refer the readers to [49] and the references therein for more details.
- The study of unique continuation problems for parabolic equations began with early works of [39, 48], followed by [43]. The study of the strong unique continuation problem was started with the work of [34], addressed to SUCP for the heat equation with bounded and time-invariant potential. This was continued with work of [7, 16, 28, 41], etc. In (3.79) of this chapter, we assume that $a \in L^\infty_{loc}(Q; \mathbb{R}^n)$ and $b \in L^\infty_{loc}(Q)$. By means of the argument in [13], the conditions on a and b can be weakened. Nevertheless, a detailed presentation of the corresponding result is beyond the scope of this book.
- An useful application of Carleman estimate for parabolic equations which is not addressed here is the derivation of the backward uniqueness of such equation (e.g., [14, 33]). Quite interestingly, it can be used to study the regularity of weak solutions to the Navier-Stokes equations for incompressible fluid. For example, it can be used to show that every weak solution to such a system is actually smooth on $(0, T) \times \mathbb{R}^3$ whenever it belongs to the space $L^\infty(0, T; L^3(\mathbb{R}^3))$ [15].
- Observability estimates (for parabolic equations, for example) on measurable sets are strongly related to time optimal control problems. We refer to [1, 40] and especially the very interesting recent book [47] for progress in this respect.
- In this chapter, we obtain Carleman estimates for parabolic equations with homogeneous boundary conditions and L^2-nonhomogeneous terms. People also considered parabolic equations with nonhomogeneous boundary conditions and H^{-1}-nonhomogeneous terms. Such kind of estimate plays an important role in obtaining the Carleman estimate for linearized Navier-Stokes equations and proving the uniqueness and stability in inverse problems of determining spatially varying coefficients in parabolic equations by overdetermining data on the lateral boundary (e.g. [4, 26]).

- Recently, Carleman estimate was generalized to stochastic parabolic equations to study controllability problems ([2, 22, 35, 44]), unique continuation problems ([32, 36, 38, 51]) and inverse problems ([37, 50]). However, this topic is far from well understood. It is an interesting new direction and some of the problems in this topic may be challenging.

References

1. Apraiz, J., Escauriaza, L., Wang, G., Zhang, C.: Observability inequalities and measurable sets. J. Eur. Math. Soc. **16**, 2433–2475 (2014)
2. Barbu, V., Răşcanu, A., Tessitore, G.: Carleman estimate and cotrollability of linear stochastic heat equatons. Appl. Math. Optim. **47**, 97–120 (2003)
3. Biccari, U., Zuazua, E.: Null controllability for a heat equation with a singular inverse-square potential involving the distance to the boundary function. J. Differ. Equ. **261**, 2809–2853 (2016)
4. Boulakia, M., Guerrero, S.: Local null controllability of a fluid-solid interaction problem in dimension 3. J. Eur. Math. Soc. **15**, 825–856 (2013)
5. Bukhgeim, A.L., Klibanov, M.V.: Global uniqueness of class of multidimensional inverse problems. Sov. Math. Dokl. **24**, 244–247 (1981)
6. Cannarsa, P., Martinez, P., Vancostenoble, J.: Global Carleman estimates for degenerate parabolic operators with applications. Mem. Am. Math. Soc. **239**, (2016)
7. Chen, X.: A strong unique continuation theorem for parabolic equations. Math. Ann. **311**, 603–630 (1998)
8. Di Cristo, M., Francini, E., Lin, C., Vessella, S., Wang, J.: Carleman estimate for second order elliptic equations with Lipschitz leading coefficients and jumps at an interface. J. Math. Pures Appl. **108**, 163–206 (2017)
9. Doubova, A, Osses, A., Puel, J.-P.: Exact controllability to trajectories for semilinear heat equations with discontinuous diffusion coefficients. ESAIM: Control Optim. Calc. Var. **8**, 621–661 (2002)
10. Doubova, A., Fernández-Cara, E., González-Burgos, M., Zuazua, E.: On the controllability of parabolic systems with a nonlinear term involving the state and the gradient. SIAM J. Control Optim. **41**, 798–819 (2002)
11. Duyckaerts, T., Zhang, X., Zuazua, E.: On the optimality of the observability inequalities for parabolic and hyperbolic systems with potentials. Ann. Inst. H. Poincaré Anal. Non Linéaire **25**, 1–41 (2008)
12. Ervedoza, S.: Control and stabilization properties for a singular heat equation with an inverse-square potential. Commun. Partial Differ. Equ. **33**, 1996–2019 (2008)
13. Escauriaza, L., Vega, L.: Carleman inequalities and the heat operator. II. Indiana Univ. Math. J. **50** (2001)
14. Escauriaza, L., Seregin, G., Šverák, V.: Backward uniqueness for parabolic equations. Arch. Ration. Mech. Anal. **169**, 147–157 (2003)
15. Escauriaza, L., Seregin, G., Šverák, V.: $L^{3,\infty}$-solutions of Navier-Stokes equations and backward uniqueness. Russ. Math. Surveys. **58**, 211–250 (2003)
16. Escauriaza, L., Fernández, F., Vessella, S.: Doubling properties of caloric functions. Appl. Anal. **85**, 205–223 (2006)
17. Fabre, C., Puel, J.P., Zuazua, E.: Approximate controllability of the semilinear heat equations. Proc. R. Soc. Edinb. **125**, 31–61 (1995)
18. Fernández-Cara, E., Zuazua, E.: Null and approximate controllability for weakly blowing up semilinear heat equations. Ann. Inst. H. Poincaré Anal. Non Linéaire **17**, 583–616 (2000)
19. Fernández-Cara, E., Zuazua, E.: The cost of approximate controllability for heat equations: the linear case. Adv. Differ. Equ. **5**, 465–514 (2000)

20. Fragnelli, G., Mugnai, D.: Carleman estimates, observability inequalities and null controllability for interior degenerate nonsmooth parabolic equations. Mem. Am. Math. Soc. **242**, (2016)

21. Fursikov, A.V., Imanuvilov, O.Y.: Controllability of Evolution Equations. Lecture Notes Series, vol. 34. Seoul National University, Seoul, Korea (1996)

22. Gao, P., Chen, M., Li, Y.: Observability estimates and null controllability for forward and backward linear stochastic Kuramoto-Sivashinsky equations. SIAM J. Control Optim. **53**, 475–500 (2015)

23. Guerrero, S.: Null controllability of some systems of two parabolic equations with one control force. SIAM J. Control Optim. **46**, 379–394 (2007)

24. Imanuvilov, O.Y., Yamamoto, M.: Lipschitz stability in inverse parabolic problems by the Carleman estimate. Inverse Probl. **14**, 1229–1245 (1998)

25. Imanuvilov, OYu.: Controllability of the parabolic equations. Sb. Math. **186**, 879–900 (1995)

26. Imanuvilov, OYu., Yamamoto, M.: Carleman inequalities for parabolic equations in a Sobolev spaces of negative order and exact controllability for semilinear parabolic equations. Publ. RIMS Kyoto Univ. **39**, 227–274 (2003)

27. Isakov, V.: Inverse Source Problems. AMS Mathematical Monographs and Surveys, Providence, vol. 34 (1990)

28. Koch, H., Tataru, D.: Carleman estimates and unique continuation for second order elliptic equations with nonsmooth coefficients. Commun. Pure Appl. Math. **54**, 339–360 (2001)

29. Le Rousseau, J., Robbiano, L.: Local and global Carleman estimates for parabolic operators with coefficients with jumps at interfaces. Invent. Math. **183**, 245–336 (2011)

30. Le Rousseau, J., Léautaud, M., Robbiano, L.: Controllability of a parabolic system with a diffuse interface. J. Eur. Math. Soc. **15**, 1485–1574 (2013)

31. Li, W., Zhang, X.: Controllability of parabolic and hyperbolic equations: Toward a unified theory. In Control Theory of Partial Differential Equations. Lecture Notes in Pure and Applied Mathematics, vol. 242, pp. 157–174. Chapan & Hall/CRC, Boca Raton, FL (2005)

32. Li, H., Lü, Q.: A quantitative boundary unique continuation for stochastic parabolic equations. J. Math. Anal. Appl. **402**, 518–526 (2013)

33. Li, L., Sverák, V.: Backward uniqueness for the heat equation in cones. Commun. Partial Differ. Equ. **37**, 1414–1429 (2012)

34. Lin, F.: A uniqueness theorem for parabolic equations. Commun. Pure Appl. Math. **43**, 127–136 (1990)

35. Liu, X.: Global Carleman estimate for stochastic parabolic equations, and its application. ESAIM Control Optim. Calc. Var. **20**, 823–839 (2014)

36. Lü, Q.: Strong unique continuation property for stochastic parabolic equations. arXiv:1701.02136

37. Lü, Q.: Carleman estimate for stochastic parabolic equations and inverse stochastic parabolic problems. Inverse Probl. **28**, 045008 (2012)

38. Lü, Q., Yin, Z.: Unique continuation for stochastic heat equations. ESAIM Control Optim. Calc. Var. **21**, 378–398 (2015)

39. Mizohata, S.: Unicité du prolongement des solutions pour quelques opérateurs différentiels paraboliques. Mem. Coll. Sci. Univ. Kyoto. Ser. A. Math. **31**, 219–239 (1958)

40. Phung, K.D., Wang, L., Zhang, C.: Bang-bang property for time optimal control of semilinear heat equation. Ann. Inst. H. Poincar Anal. Non Linéaire **31**, 477–499 (2014)

41. Poon, C.: Unique continuation for parabolic equations. Commun. Partial Differ. Equ. **21**, 521–539 (1996)

42. Rosier, L., Zhang, B.-Y.: Null controllability of the complex Ginzburg-Landau equation. Ann. Inst. H. Poincaré Anal. Non Linéaire. **26**, 649–673 (2009)

43. Saut, J.C., Scheurer, B.: Unique continuation for some evolution equations. J. Differ. Equ. **66**, 118–139 (1987)

44. Tang, S., Zhang, X.: Null controllability for forward and backward stochastic parabolic equations. SIAM J. Control Optim. **48**, 2191–2216 (2009)

45. Vancostenoble, J., Zuazua, E.: Null controllability for the heat equation with singular inverse-square potentials. J. Funct. Anal. **254**, 1864–1902 (2008)
46. Vessella, S.: Carleman estimates, optimal three cylinder inequality, and unique continuation properties for solutions to parabolic equations. Commun. Partial Differ. Equ. **28**, 637–676 (2003)
47. Wang, G., Wang, L., Xu, Y., Zhang, Y.: Time Optimal Control of Evolution Equations. Progress in Nonlinear Differential Equations and their Applications. Subseries in Control, vol. 92. Birkhäuser/Springer, Cham (2018)
48. Yamabe, H.: A unique continuation theorem of a diffusion equation. Ann. Math. **69**, 462–466 (1959)
49. Yamamoto, M.: Carleman estimates for parabolic equations and applications. Inverse Probl. **25**, 123013 (2009)
50. Yuan, G.: Conditional stability in determination of initial data for stochastic parabolic equations. Inverse Probl. **33**, 035014 (2017)
51. Zhang, X.: Unique continuation for stochastic parabolic equations. Differ. Integral Equ. **21**, 81–93 (2008)
52. Zuazua, E.: Controllability and observability of partial differential equations: some results and open problems. In: Handbook of Differential Equations: Evolutionary Differential Equations, vol. 3, pp. 527–621. Elsevier Science, Amsterdam (2006)

Chapter 4
Carleman Estimates for Second Order Hyperbolic Operators and Applications, a Unified Approach

Abstract In this chapter, we establish three Carleman estimates for second order hyperbolic operators. The first one is Theorem 4.1, which is used to solve an inverse hyperbolic problem. The second one is Theorem 4.2, a Carleman estimate in H^1-norm, and based on it, we further derive the third Carleman estimate in L^2-norm (see Theorem 4.3). As the applications of the later, we obtain the exact controllability of semilinear hyperbolic equations and the exponential decay of locally damped hyperbolic equations.

Keywords Carleman estimate · Second order hyperbolic operator · Exact controllability · Exponential decay · Inverse hyperbolic problem

Throughout this chapter, we need the following assumption.

Condition 4.1 *The functions* $h^{jk}(\cdot) \in C^2(\overline{\Omega}; \mathbb{R})$ *(for* $j, k = 1, \cdots, n$*) satisfy*

$$h^{jk}(x) = h^{kj}(x), \qquad \forall\, x \in \overline{\Omega}, \tag{4.1}$$

and for some constant $h_0 > 0$,

$$\sum_{j,k=1}^{n} h^{jk}(x)\xi^j\overline{\xi}^k \geq h_0|\xi|^2, \qquad \forall\, (x, \xi^1, \cdots, \xi^n) \in \overline{\Omega} \times \mathbb{C}^n. \tag{4.2}$$

Remark 4.1 Condition 4.1 is very similar to Condition 2.1 in Chap. 2. The only difference is the regularity of $h^{jk}(\cdot)$ (for $j, k = 1, \cdots, n$).

4.1 Carleman Estimates for Second Order Hyperbolic Operators

We first establish the following pointwise estimate, which is an immediate consequence of Lemma 2.1.

© The Author(s), under exclusive license to Springer Nature Switzerland AG 2019
X. Fu et al., *Carleman Estimates for Second Order Partial Differential Operators and Applications*, SpringerBriefs in Mathematics,
https://doi.org/10.1007/978-3-030-29530-1_4

Corollary 4.1 *Let* $z \in C^2(\mathbb{R}^{1+n}; \mathbb{R})$, $\ell \in C^3(\mathbb{R}^{1+n}; \mathbb{R})$ *and* $\Psi \in C^1(\mathbb{R}^n; \mathbb{R})$. *Set* $\theta = e^\ell$ *and* $v = \theta z$. *Then*

$$
e^{2\lambda\phi} \left| z_{tt} - \sum_{j,k=1}^n \left(h^{jk} z_{x_j} \right)_{x_k} \right|^2 + 2\mathrm{div}\, V + 2M_t
$$
$$
\geq 2 \left[\ell_{tt} + \sum_{j,k=1}^n \left(h^{jk} \ell_{x_j} \right)_{x_k} + \Psi \right] v_t^2 - 8 \sum_{j,k=1}^n h^{jk} \ell_{tx_j} v_{x_k} v_t \tag{4.3}
$$
$$
+ 2 \sum_{j,k=1}^n c^{jk} v_{x_j} v_{x_k} - 2 \sum_{j,k=1}^n h^{jk} \Psi_{x_j} v v_{x_k} + B v^2,
$$

where

$$
\begin{cases}
A = \displaystyle\sum_{j,k=1}^n \left(h^{jk} \ell_{x_j} \ell_{x_k} - h^{jk}_{x_j} \ell_{x_k} - h^{jk} \ell_{x_j x_k} \right) - \ell_t^2 + \ell_{tt} - \Psi, \\[2mm]
c^{jk} = \displaystyle\sum_{j',k'=1}^n \left[2 h^{jk'} (h^{j'k} \ell_{x_{j'}})_{x_{k'}} - \left(h^{jk} h^{j'k'} \ell_{x_{j'}} \right)_{x_{k'}} \right] + h^{jk} (\ell_{tt} - \Psi), \\[2mm]
B = 2 \left[A\Psi - (A\ell_t)_t + \displaystyle\sum_{j,k=1}^n \left(A h^{jk} \ell_{x_j} \right)_{x_k} \right],
\end{cases} \tag{4.4}
$$

and

$$
\begin{cases}
V = \left[V^1, \cdots, V^k, \cdots, V^n \right], \\[2mm]
V^k = 2 \displaystyle\sum_{j,j',k'=1}^n h^{jk} h^{j'k'} \ell_{x_{j'}} v_{x_j} v_{x_{k'}} + \sum_{j=1}^n h^{jk} A \ell_{x_j} v^2 - \Psi v \sum_{j=1}^n h^{jk} v_{x_j} \\[2mm]
\quad - \displaystyle\sum_{j,j',k'=1}^n h^{jk} h^{j'k'} \ell_{x_j} v_{x_{j'}} v_{x_{k'}} - 2\ell_t v_t \sum_{j=1}^n h^{jk} v_{x_j} + \sum_{j=1}^n h^{jk} \ell_{x_j} v_t^2, \\[2mm]
M = \ell_t \left(v_t^2 + \displaystyle\sum_{j,k=1}^n h^{jk} v_{x_j} v_{x_k} \right) - 2 \sum_{j,k=1}^n h^{jk} \ell_{x_j} v_{x_k} v_t + \Psi v v_t - A \ell_t v^2.
\end{cases} \tag{4.5}
$$

Proof Using Lemma 2.1 with $\Phi = 0$, $m = 1 + n$ and

$$
(a^{jk})_{m \times m} = \begin{pmatrix} -1 & 0 \\ 0 & (h^{jk})_{n \times n} \end{pmatrix}.
$$

We immediately obtain (4.3).

Now, let us consider the following hyperbolic equation:

$$\begin{cases} z_{tt} - \sum_{j,k=1}^{n} (h^{jk} z_{x_j})_{x_k} = F & \text{in } Q, \\ z = 0 & \text{on } \Sigma, \\ z(0) = z_0, \ z_t(0) = z_1 & \text{in } \Omega, \end{cases} \tag{4.6}$$

where $F \in L^2(Q)$ and (z_0, z_1) belongs to some suitable space (which will be given later). In case of $(z_0, z_1) \in H_0^1(\Omega) \times L^2(\Omega)$, we establish two Carleman estimates in H^1-norm. In case of $(z_0, z_1) \in L^2(\Omega) \times H^{-1}(\Omega)$, we obtain a Carleman estimate in L^2-norm.

We need more assumptions on $(h^{jk})_{1 \le j,k \le n}$ as follows:

Condition 4.2 *There is a positive function $\psi(\cdot) \in C^3(\overline{\Omega})$ such that $\min_{x \in \overline{\Omega}} |\nabla \psi(x)| > 0$ and that, for some constant $\mu_0 > 0$,*

$$\sum_{j,k=1}^{n} \sum_{j',k'=1}^{n} \left[2h^{jk'} (h^{j'k} \psi_{x_{j'}})_{x_{k'}} - h_{x_{k'}}^{jk} h^{j'k'} \psi_{x_{j'}} \right] \xi^j \xi^k \ge \mu_0 \sum_{j,k=1}^{n} h^{jk} \xi^j \xi^k, \tag{4.7}$$
$$\forall (x, \xi^1, \cdots, \xi^n) \in \overline{\Omega} \times \mathbb{R}^n.$$

Remark 4.2 Condition 4.2 is a sufficient condition to establish the Carleman estimate for the hyperbolic operator $\partial_t^2 - \sum_{j,k=1}^{n} \partial_{x_k} (h^{jk} \partial_{x_j})$. If $(h^{jk})_{1 \le j,k \le n} = I_n$ (the $n \times n$ identity matrix), then $\psi(x) = |x - x_0|^2$ satisfies Condition 4.2 with $\mu_0 = 4$ and (4.7) holds as an equality, where x_0 is any given point in $\mathbb{R}^n \setminus \overline{\Omega}$. We refer to [12, 24] for examples for which Condition 4.2 is satisfied and for more explanation on this condition.

Put

$$\Gamma_0 \overset{\triangle}{=} \left\{ x \in \Gamma : \sum_{j,k=1}^{n} h^{jk}(x) \psi_{x_j}(x) \nu^k(x) > 0 \right\}. \tag{4.8}$$

Also, for any $\delta > 0$, write

$$\mathcal{O}_\delta(\Gamma_0) \overset{\triangle}{=} \{ x \in \mathbb{R}^n : \text{dist}(x, \Gamma_0) < \delta \}, \quad \omega \overset{\triangle}{=} \mathcal{O}_\delta(\Gamma_0) \bigcap \Omega. \tag{4.9}$$

Remark 4.3 If $(h^{jk})_{1 \le j,k \le n} = I_n$, then the set defined by (4.8) is specialized as

$$\Gamma_0 = \{ x \in \Gamma : (x - x_0) \cdot \nu(x) > 0 \},$$

which coincides in the usual star-shaped sub-boundary of Γ [23].

For some constants $c_0, c_1 \in (0, 1)$ (which will be given later) and parameter $\lambda > 0$, in the rest of this chapter we shall choose the weight function θ and the auxiliary function Ψ (appeared in Corollary 4.1) as follows:

$$\begin{cases} \theta(t,x) = e^{\ell(t,x)}, \quad \ell(t,x) = \lambda\phi(t,x), \\ \phi(t,x) = \psi(x) - c_1(t - T/2)^2, \\ \Psi(x) = -\lambda\Big[\sum_{j,k=1}^{n} (h^{jk}\psi_{x_j})_{x_k} - 2c_1 - c_0 \Big], \end{cases} \tag{4.10}$$

where $\psi(\cdot)$ is given by Condition 4.2.

4.1.1 Carleman Estimate in H^1-Norm

This subsection is devoted to deriving a global Carleman estimate for the hyperbolic equation (4.6) in H^1-norm.

It is easy to see that, if $\psi(\cdot) \in C^2(\overline{\Omega})$ satisfies Condition 4.2, then for any given constants $a \geq 1$ and $b \in \mathbb{R}$, the function

$$\hat{\psi} = \hat{\psi}(x) \stackrel{\triangle}{=} a\psi(x) + b \tag{4.11}$$

(scaling and translating $\psi(x)$) still satisfies Condition 4.2 with μ_0 replaced by $a\mu_0$; meanwhile, the scaling and translating $\psi(x)$ does not change the set Γ_0. Hence, without loss of generality, we may assume that

$$\begin{cases} \text{Condition 4.1.1 holds with } \mu_0 \geq 4, \\ \dfrac{1}{4} \sum_{j,k=1}^{n} h^{jk}(x)\psi_{x_j}(x)\psi_{x_k}(x) \geq \max_{x\in\Omega}\psi(x) \geq \min_{x\in\Omega}\psi(x) > 0, \quad \forall\, x \in \overline{\Omega}. \end{cases} \tag{4.12}$$

Let

$$R_1 \stackrel{\triangle}{=} \max_{x\in\Omega}\sqrt{\psi(x)}, \quad T_0 \stackrel{\triangle}{=} 2 \inf\big\{ R_1 \;:\; \psi(\cdot)\text{satisfies (4.12)}\big\}. \tag{4.13}$$

We have the following boundary global Carleman estimate for the equation (4.6).

Theorem 4.1 *Let Condition 4.2 hold and Γ_0 be given in (4.8). Then there is a positive constant λ_0 such that for any $T > T_0$ and $\lambda \geq \lambda_0$, there exist $c_2 > 0$ and $C > 0$ such that every solution $z \in H^1(Q)$ to (4.6) satisfies that*

$$\int_Q \theta^2\big[\lambda(z_t^2 + |\nabla z|^2) + \lambda^3 z^2\big]dxdt$$
$$\leq C\Big(\int_Q \theta^2|F|^2 dxdt + \lambda^3 e^{-c_2\lambda}E(0) + \lambda \int_0^T \int_{\Gamma_0} \theta^2 \Big|\frac{\partial z}{\partial \nu}\Big|^2 dxdt \Big). \tag{4.14}$$

Proof Let us divide the proof into two steps.

Step 1. Recalling (4.10) for the definitions of Ψ and ℓ, it is clear that

$$2\left[\ell_{tt} + \sum_{j,k=1}^{n} (h^{jk}\ell_{x_j})_{x_k} + \Psi\right] = -4\lambda c_1 + 2\lambda \sum_{j,k=1}^{n} (h^{jk}\psi_{x_j})_{x_k} + 2\Psi = 2\lambda c_0.$$

(4.15)

Further, by (4.4) and (4.10), we have

$$\sum_{j,k=1}^{n} c^{jk} v_{x_j} v_{x_k}$$

$$= \sum_{j,k=1}^{n} \left\{ \sum_{j',k'=1}^{n} \left[2h^{jk'}(h^{j'k}\ell_{x_{j'}})_{x_{k'}} - (h^{jk}h^{j'k'}\ell_{x_{j'}})_{x_{k'}}\right] + (\ell_{tt} - \Psi)h^{jk} \right\} v_{x_j} v_{x_k}$$

$$= \sum_{j,k=1}^{n} \left[-2\lambda c_1 h^{jk} + \sum_{j',k'=1}^{n} \lambda h^{jk}(h^{j'k'}\psi_{x_{j'}})_{x_{k'}} - 2\lambda c_1 h^{jk} - c_0\lambda h^{jk} \right. \qquad (4.16)$$

$$\left. +2\lambda \sum_{j',k'=1}^{n} h^{jk'}(h^{j'k}\psi_{x_{j'}})_{x_{k'}} - \lambda \sum_{j',k'=1}^{n} (h^{jk}h^{j'k'}\psi_{x_{j'}})_{x_{k'}} \right] v_{x_j} v_{x_k}$$

$$\geq \lambda\mu_0 \sum_{j,k=1}^{n} h^{jk} v_{x_j} v_{x_k} - (4c_1 + c_0)\lambda \sum_{j,k=1}^{n} h^{jk} v_{x_j} v_{x_k}$$

$$= \lambda(\mu_0 - 4c_1 - c_0) \sum_{j,k=1}^{n} h^{jk} v_{x_j} v_{x_k}.$$

Next, by using (4.4) and (4.10) again, we obtain that

$$A = \sum_{j,k=1}^{n} (h^{jk}\ell_{x_j}\ell_{x_k} - h^{jk}_{x_j}\ell_{x_k} - h^{ij}\ell_{x_j x_k}) - \ell_t^2 + \ell_{tt} + \Psi$$

$$= \lambda^2\left[\sum_{j,k=1}^{n} h^{jk}\psi_{x_j}\psi_{x_k} - c_1^2(2t - T)^2 \right] + O(\lambda). \qquad (4.17)$$

It follows from (4.4), (4.10) and (4.17) that

$$B = 2\left[A\Psi - (A\ell_t)_t + \sum_{j,k=1}^{n} \left(Ah^{jk}\ell_{x_j} \right)_{x_k} \right]$$

$$= 2A\left[\Psi - \ell_{tt} + \sum_{j,k=1}^{n} \left(h^{jk}\ell_{x_j} \right)_{x_k} \right] + 2\left(\sum_{j,k=1}^{n} h^{jk}\ell_{x_j} A_{x_k} - A_t\ell_t \right)$$

$$= (4c_1 + c_0)\lambda A + 2\left(\sum_{j,k=1}^{n} h^{jk}\ell_{x_j} A_{x_k} - A_t\ell_t \right) \qquad (4.18)$$

$$= 2\lambda^3\left[(4c_1 + c_0) \sum_{j,k=1}^{n} h^{jk}\psi_{x_j}\psi_{x_k} + \sum_{j,k=1}^{n} h^{jk}\psi_{x_j}\left(\sum_{j',k'=1}^{n} h^{j'k'}\psi_{x_{j'}}\psi_{x_{k'}} \right)_{x_k} \right.$$

$$\left. -(8c_1 + c_0)c_1^2(2t - T)^2 \right] + O(\lambda^2).$$

Further, recalling that ψ satisfies Condition 4.2, and noting $h^{j'k'} = h^{k'j'}$ for $1 \leq j', k' \leq n$, we find that

$$
\mu_0 \sum_{j,k=1}^{n} h^{jk} \psi_{x_j} \psi_{x_k}
$$

$$
\leq \sum_{j,k,j',k'=1}^{n} \left[2h^{jk'}(h^{j'k}\psi_{x_{j'}})_{x_{k'}} - h^{jk}_{x_{k'}} h^{j'k'} \psi_{x_{j'}} \right] \psi_{x_j} \psi_{x_k}
$$

$$
= \sum_{j,k,j',k'} \left(2h^{jk'} h^{j'k}_{x_{k'}} \psi_{x_{j'}} \psi_{x_j} \psi_{x_k} + 2h^{jk'} h^{j'k} \psi_{x_{j'k'}} \psi_{x_j} \psi_{x_k} - h^{jk}_{x_{k'}} h^{j'k'} \psi_{x_{j'}} \psi_{x_j} \psi_{x_k} \right)
$$

$$
= \sum_{j,k,j',k'=1}^{n} \left(h^{jk'} h^{j'k}_{x_{k'}} \psi_{x_{j'}} \psi_{x_j} \psi_{x_k} + 2h^{jk'} h^{j'k} \psi_{x_{j'k'}} \psi_{x_j} \psi_{x_k} \right) \tag{4.19}
$$

$$
= \sum_{j,k,j',k'=1}^{n} \left(h^{jk} h^{j'k'}_{x_j} \psi_{x_{j'}} \psi_{x_j} \psi_{x_{k'}} + 2h^{jk} h^{j'k'} \psi_{x_{j'k}} \psi_{x_j} \psi_{x_{k'}} \right)
$$

$$
= \sum_{j,k,j',k'=1}^{n} \left(h^{jk} h^{j'k'}_{x_k} \psi_{x_{j'}} \psi_{x_j} \psi_{x_{k'}} + h^{jk} h^{j'k'} \psi_{x_{j'k}} \psi_{x_j} \psi_{x_{k'}} + h^{jk} h^{k'j'} \psi_{x_{k'k}} \psi_{x_j} \psi_{x_{k'}} \right)
$$

$$
= \sum_{j,k=1}^{n} h^{jk} \psi_{x_j} \left(\sum_{j',k'=1}^{n} h^{j'k'} \psi_{x_{j'}} \psi_{x_{k'}} \right)_{x_k}.
$$

Let us now give the choices of c_0 and c_1 in (4.10). Fix $T > T_0$. Clearly, $T > 2R_1$. Thus, one can choose a constant $c_1 \in (0,1)$ such that

$$
\left(\frac{2R_1}{T} \right)^2 < c_1 < \frac{2R_1}{T}. \tag{4.20}
$$

Noting that $\mu_0 > 4$, for a fixed c_1, one can find a constant $c_0 \in (0,1)$ such that

$$
0 < c_0 < \min \left\{ 1, \frac{\mu_0 - 4c_1}{2} \right\}. \tag{4.21}
$$

By (4.18)–(4.21), noting that $\mu_0 > 4c_1 + c_0$, we find that,

$$
\begin{aligned}
B &\geq 2\lambda^3 (8c_1 + c_0) \left[\sum_{j,k=1}^{n} h^{jk} \psi_{x_j} \psi_{x_k} - c_1^2 (2t - T)^2 \right] + O(\lambda^2) \\
&\geq 16\lambda^3 c_1 (4R_1^2 - c_1^2 T^2) + O(\lambda^2).
\end{aligned} \tag{4.22}
$$

By (4.3), (4.15), (4.16) and (4.20)–(4.22), there exist $\lambda_1 > 0$ and $c^* > 0$, such that for any $\lambda \geq \lambda_1$,

$$2\left[\ell_{tt} + \sum_{j,k=1}^{n} (h^{jk}\ell_{x_j})_{x_k} + \Psi\right]v_t^2 - 8\sum_{j,k=1}^{n} h^{jk}\ell_{x_j t}v_{x_k}v_t$$

$$+2\sum_{j,k=1}^{n} c^{jk}v_{x_j}v_{x_k} - 2\sum_{j,k=1}^{n} h^{jk}\Psi_{x_j}vv_{x_i} + Bv^2 \tag{4.23}$$

$$\geq c^*\lambda\left(v_t^2 + |\nabla v|^2 + \lambda^2 v^2\right).$$

Integrating (4.23) in Q and noting that $v = 0$ on Σ (which follows from $v = \theta z$ and $z = 0$ on Σ), by (4.3) and (4.6), we obtain that

$$c^*\lambda \int_{Q} \left(v_t^2 + |\nabla v|^2 + \lambda^2 v^2\right)dxdt$$

$$\leq \int_{Q} \theta^2 |F|^2 dxdt + 2\int_{Q} M_t dxdt + 2\lambda s_0 \int_{0}^{T}\int_{\Gamma_0} \sum_{j,k=1}^{n} h^{jk}v^j v^k \theta^2 \left|\frac{\partial z}{\partial \nu}\right|^2 d\Sigma. \tag{4.24}$$

Here $s_0 = \max_{x\in\Gamma} \sum_{j,k=1}^{n} h^{jk}\psi_{x_j}v^k$ and we have used the following equality:

$$\int_{\Sigma} \sum_{j,k,j',k'=1}^{n} \left(2h^{jk}h^{j'k'}\ell_{x_{j'}}v_{x_j}v_{x_{k'}} - h^{jk}h^{j'k'}\ell_{x_j}v_{x_{j'}}v_{x_{k'}}\right)v^k d\Sigma$$

$$= \lambda \int_{\Sigma} \sum_{j,k=1}^{n} h^{jk}v^j v^k \sum_{j'k'=1}^{n} h^{j'k'}\psi_{x_{j'}}v^{k'}\left|\frac{\partial v}{\partial \nu}\right|^2 d\Sigma.$$

Step 2. In this step, we handle the term $\int_{Q} M_t dxdt$. By (4.10), (4.13) and (4.20), we have

$$\phi(0, x) = \phi(T, x) < R_1^2 - c_1 T^2/4 < 0, \qquad \forall\, x \in \Omega. \tag{4.25}$$

Hence, there exist $T_* \in (0, T)$, close to 0, and $T^* \in (0, T)$, close to T, such that

$$\phi(t, x) \leq R_1^2/2 - c_1 T^2/8 < 0, \qquad \forall\, (t, x) \in \left((0, T_*) \cup (T^*, T)\right) \times \Omega. \tag{4.26}$$

Let

$$E(t) = \frac{1}{2}\int_{\Omega} (|\nabla z|^2 + |z_t|^2)dx.$$

Recalling (4.5) for the definition of M, by (4.10) and (4.26), and recalling that $v = \theta z$, we have

$$\left|\int_{Q} M_t dxdt\right| = \left|\int_{\Omega} \left(M(T, x) - M(0, x)\right)dx\right|$$

$$\leq C\lambda^3 e^{(R_1^2 - c_1 T^2/4)\lambda}\left(E(0) + E(T)\right) \tag{4.27}$$

$$\leq C\lambda^3 e^{-c_2\lambda}E(0),$$

where $c_2 = c_1 T^2/4 - R_1^2 > 0$.

Finally, by (4.24) and (4.27), and noting that $v = \theta z$, we complete the proof.

In what follows, set

$$\kappa_1 = \max_{x \in \bar\Omega} \sum_{j,k=1}^n h^{jk} \psi_{x_j} \psi_{x_k}, \qquad s_0 = \max_{x \in \Gamma} \sum_{j,k=1}^n h^{jk} \psi_{x_j} v^k.$$

Put

$$T_1 \overset{\triangle}{=} \max \left[2\sqrt{\kappa_1}, 1 + \frac{24\sqrt{n}s_0}{\min\{1, h_0\}} \left(1 + \frac{1}{h_0^{3/2}} \sum_{j,k=1}^n |h^{jk}|_{C(\bar\Omega)} + \frac{1}{h_0} \right) \right], \qquad (4.28)$$

where h_0 is the constant appeared in (4.2).

By scaling and translating of $\psi(\cdot)$ if necessary, in what follows, we may assume that $\psi(\cdot)$ and μ_0 satisfy that

$$\kappa_0 \overset{\triangle}{=} \min_{x \in \bar\Omega} \sum_{j,k=1}^n h^{jk} \psi_{x_j} \psi_{x_k} \geq \max_{x \in \bar\Omega} \psi(x) \quad \text{and} \quad \mu_0 > \frac{9T_1^2}{\kappa_0}. \qquad (4.29)$$

We have the following Carleman estimate for the Eq. (4.6) in H^1-norm.

Theorem 4.2 *Let Condition 4.2 hold, and ω and T_1 be given in (4.9) and (4.28), respectively. Then there is a positive constant λ_0, such that for all $T > T_1$ and $\lambda \geq \lambda_0$, every solution $z \in H^1(Q)$ to (4.6) satisfies that*

$$\int_Q \theta^2 \left[\lambda(z_t^2 + |\nabla z|^2) + \lambda^3 z^2 \right] dx dt$$
$$\leq C \left[\int_Q \theta^2 F^2 dx dt + \lambda^2 \int_0^T \int_\omega \theta^2 (z_t^2 + \lambda^2 z^2) dx dt \right]. \qquad (4.30)$$

Proof Note that $T > T_1 \geq T_0$. By (4.29), it is easy to see that $\frac{T_0}{T} < \frac{\sqrt{\mu_0 \kappa_0}}{3T}$. Consequently, we can choose a constant

$$c_1 \in \left(\frac{T_0}{T}, \min\left\{ 1, \frac{\sqrt{\mu_0 \kappa_0}}{3T} \right\} \right).$$

Clearly, (4.24) still holds. Therefore, we only need to estimate its second and third term. The procedure is divided into two steps.

Step 1. Estimation of "the spatial boundary term". We choose functions $\rho_0 \in C^1(\bar\Omega; \mathbb{R}^n)$ and $\rho_1 \in C^2(\bar\Omega; [0, 1])$, such that $\rho_0 = v$ on Γ, and

$$\begin{cases} \rho_1(x) \equiv 1, & x \in \mathcal{O}_{\delta/3}(\Gamma_0) \cap \Omega, \\ \rho_1(x) \equiv 0, & x \in \Omega \setminus \mathcal{O}_{\delta/2}(\Gamma_0). \end{cases}$$

Let $\eta_1 = \rho_0\rho_1\theta$. Then, a direct computation shows that

$$2\Big[z_{tt} - \sum_{j,k=1}^{n}(h^{jk}z_{x_j})_{x_k}\Big]\eta_1 \cdot \nabla z$$

$$= 2(z_t\eta_1 \cdot \nabla z)_t - 2z_t\partial_t\eta_1 \cdot \nabla z - \nabla \cdot (\eta_1 z_t^2) + (\nabla \cdot \eta_1)z_t^2 - 2\sum_{j,k=1}^{n}\Big[h^{jk}z_{x_j}(\eta_1 \cdot \nabla z)\Big]_{x_k}$$

$$+2\sum_{j,k,l=1}^{n}h^{jk}z_{x_j}z_{x_l}\frac{\partial\eta_1^l}{\partial x_k} + 2\sum_{j,k,l=1}^{n}h^{jk}z_{x_j}\eta_1^l z_{x_l x_k}. \tag{4.31}$$

Noting that $h^{jk} = h^{kj}$ $(j, k = 1, 2, \cdots, n)$, we have

$$2\sum_{j,k,l=1}^{n}h^{jk}z_{x_j}\eta_1^l z_{x_l x_k} = \sum_{j,k,l=1}^{n}\Big(h^{jk}z_{x_j}\eta_1^l z_{x_l x_k} + h^{kj}z_{x_k}\eta_1^l z_{x_l x_j}\Big)$$

$$= \sum_{j,k,l=1}^{n}h^{jk}\eta_1^l(z_{x_k}z_{x_j})_{x_l} = \sum_{j,k,l=1}^{n}h^{jl}\eta_1^k(z_{x_l}z_{x_j})_{x_k} \tag{4.32}$$

$$= \sum_{j,k,l=1}^{n}(h^{jl}\eta_1^k z_{x_l}z_{x_j})_{x_k} - \sum_{j,k,l=1}^{n}(h^{jl}\eta_1^k)_{x_k}z_{x_l}z_{x_j}.$$

It follows from (4.6), (4.31) and (4.32) that

$$-\sum_{k=1}^{n}\Big[2(\eta_1 \cdot \nabla z)\sum_{j=1}^{n}h^{jk}z_{x_j} + \eta_1^k\Big(z_t^2 - \sum_{j,l=1}^{n}h^{jl}z_{x_j}z_{x_l}\Big)\Big]_{x_k}$$

$$= 2\Big[F\eta_1 \cdot \nabla z - (z_t\eta_1 \cdot \nabla z)_t + z_t\partial_t\eta_1 \cdot \nabla z - \sum_{j,k,l=1}^{n}h^{jk}z_{x_j}z_{x_l}\frac{\partial\eta_1^l}{\partial x_k}\Big] \tag{4.33}$$

$$-(\nabla \cdot \eta_1)z_t^2 + \sum_{j,k=1}^{n}z_{x_j}z_{x_k}\nu \cdot (h^{jk}\eta_1).$$

Integrating (4.33) in Q and noting that $z = 0$ on Σ, we obtain that

$$\int_\Sigma \sum_{j,k=1}^{n}h^{jk}\nu^j\nu^k\eta_1\theta^2\Big|\frac{\partial z}{\partial\nu}\Big|^2 dxdt$$

$$= -\int_Q \Big\{2\big[F\eta_1 \cdot \nabla z - (z_t\eta_1 \cdot \nabla z)_t + z_t\partial_t\eta_1 \cdot \nabla z - (\nabla \cdot \eta_1)z_t^2\big]$$

$$- \sum_{j,k,l=1}^{n}h^{jk}z_{x_j}z_{x_l}\frac{\partial\eta_1^l}{\partial x_k} + \sum_{j,k=1}^{n}z_{x_j}z_{x_k}\nabla \cdot (h^{jk}\eta_1)\Big\}dxdt.$$

Consequently,

$$\int_{\Sigma} \sum_{j,k=1}^{n} h^{jk} v^j v^k \eta_1 \theta^2 \left| \frac{\partial z}{\partial v} \right|^2 dxdt$$

$$\leq \frac{C}{\lambda} \int_Q \theta^2 |F|^2 dxdt + \lambda \int_0^T \int_{\mathscr{O}_{\delta/2}(\Gamma_0) \cap \Omega} \theta^2 |\nabla z|^2 dxdt + 2 \int_\Omega z_t \eta_1 \cdot \nabla z dx \Big|_0^T$$

$$+ C\lambda \int_0^T \int_{\mathscr{O}_{\delta/2}(\Gamma_0) \cap \Omega} \theta^2 z_t^2 dxdt + 4\sqrt{n}\lambda \int_0^T \int_{\mathscr{O}_{\delta/2}(\Gamma_0) \cap \Omega} \theta^2 c_1 T |z_t| |\nabla z| dxdt$$

$$+ \left(6\sqrt{n}\lambda |\nabla \psi| \sum_{j,k=1}^{n} |h^{jk}|_{C(\overline{\Omega})} + C \right) \int_0^T \int_{\mathscr{O}_{\delta/2}(\Gamma_0) \cap \Omega} \theta^2 |\nabla z|^2 dxdt$$

$$\leq C \left(\frac{1}{\lambda} \int_Q \theta^2 |F|^2 dxdt + \lambda \int_0^T \int_{\mathscr{O}_{\delta/2}(\Gamma_0) \cap \Omega} \theta^2 z_t^2 dxdt \right) + 2 \int_\Omega z_t \eta_1 \cdot \nabla z dx \Big|_0^T$$

$$+ 6\sqrt{n}\lambda \left(|\nabla \psi| \sum_{j,k=1}^{n} |h^{jk}|_{C(\overline{\Omega})} + 1 \right) \int_0^T \int_{\mathscr{O}_{\delta/2}(\Gamma_0) \cap \Omega} \theta^2 |\nabla z|^2 dxdt.$$

This, together with (4.24), implies that

$$c^* \lambda \int_Q (v_t^2 + |\nabla v|^2 + \lambda^2 v^2) dxdt$$

$$\leq C \left(\int_Q \theta^2 |F|^2 dxdt + \lambda^2 \int_0^T \int_\omega \theta^2 z_t^2 dxdt \right) + 4\lambda s_0 \int_\Omega z_t \eta_1 \cdot \nabla z dx \Big|_0^T + 2 \int_\Omega \widetilde{M} dx \Big|_0^T$$

$$+ 12\sqrt{n}\lambda^2 s_0 \left(|\nabla \psi| \sum_{j,k=1}^{n} |h^{jk}|_{C(\overline{\Omega})} + 1 \right) \int_0^T \int_{\mathscr{O}_{\delta/2}(\Gamma_0) \cap \Omega} \theta^2 |\nabla z|^2 dxdt, \qquad (4.34)$$

where $\widetilde{M} = M + 2\lambda s_0 z_t \eta_1 \cdot \nabla z$.

Next, we analyze the last term in (4.34). Put $\eta_2(t, x) = \rho_2^2 \theta^2$, where $\rho_2 \in C^2(\overline{\Omega}; [0, 1])$ satisfies that

$$\begin{cases} \rho_2(x) \equiv 1, & x \in \mathscr{O}_{\delta/2}(\Gamma_0) \cap \Omega, \\ \rho_2(x) \equiv 0, & x \in \Omega \setminus \omega. \end{cases}$$

It follows from (4.6) that

$$\int_Q \eta_2 z F dxdt = \int_Q \eta_2 z \left[z_{tt} - \sum_{j,k=1}^{n} (h^{jk} z_{x_j})_{x_k} \right] dxdt$$

$$= \int_Q (\eta_2 z z_t)_t dxdt - \int_Q z_t (\partial_t \eta_2 z + \eta_2 z_t) dxdt$$

$$+ \int_Q \eta_2 \sum_{j,k=1}^{n} h^{jk} z_{x_j} z_{x_k} dxdt + \int_Q z \sum_{j,k=1}^{n} h^{jk} z_{x_j} \partial_{x_k} \eta_2 dxdt,$$

which implies

$$\int_0^T \int_{\mathscr{O}_{\delta/2}(\Gamma_0) \cap \Omega} \theta^2 |\nabla z|^2 \, dx dt$$

$$\leq C \left[\frac{1}{\lambda^2} \int_Q \theta^2 |F|^2 \, dx dt + \int_0^T \int_\omega \theta^2 (\lambda^2 z^2 + z_t^2) \, dx dt \right] - \frac{1}{h_0} \int_Q (\eta_2 z z_t)_t \, dx dt.$$

This, together with (4.34), implies that

$$c^* \lambda \int_Q (v_t^2 + |\nabla v|^2 + \lambda^2 v^2) \, dx dt$$

$$\leq C \int_Q \theta^2 |F|^2 \, dx dt + \int_\Omega \overline{M} \, dx \Big|_0^T + C\lambda^2 \int_0^T \int_\omega \theta^2 (z_t^2 + \lambda^2 z^2) \, dx dt, \tag{4.35}$$

where

$$\overline{M} = M + 2 s_0 \lambda z_t \eta_1 \cdot \nabla z - \frac{1}{h_0} \left[12 \sqrt{n} \lambda^2 s_0 \Big(|\nabla \psi| \sum_{j,k=1}^n |h^{jk}|_{C(\overline{\Omega})} + 1 \Big) \right] \eta_2 z z_t.$$

Step 2. Estimation of "the time boundary term". In this step, we estimate $\overline{M}(0, x)$ and $\overline{M}(T, x)$, respectively. By (4.29) and the definition of M in (4.5), we have that

$$M(0, x) \geq \left[\lambda c_1 T \Big(v_t^2 + \sum_{j,k=1}^n h^{jk} v_{x_j} v_{x_k} \Big) - 2\lambda \sum_{j,k=1}^n h^{jk} \psi_{x_j} v_{x_k} v_t \right]\Big|_{t=0}$$

$$+ \left[O(\lambda^2) v^2 - v_t^2 + c_1 T \lambda^3 \Big(c_1^2 T^2 - \sum_{j,k=1}^n h^{jk} \psi_{x_j} \psi_{x_k} \Big) v^2 \right]\Big|_{t=0}$$

$$\geq \lambda \left[c_1 T - \Big(\sum_{j,k=1}^n h^{jk} \psi_{x_j} \psi_{x_k} \Big)^{\frac{1}{2}} \right] \Big(v_t^2 + \sum_{j,k=1}^n h^{jk} v_{x_j} v_{x_k} \Big)\Big|_{t=0}$$

$$+ \left[O(\lambda^2) v^2 - v_t^2 + c_1 T \lambda^3 \Big(c_1^2 T^2 - \sum_{j,k=1}^n h^{jk} \psi_{x_j} \psi_{x_k} \Big) v^2 \right]\Big|_{t=0}.$$

Noting that by (4.28) and $c_1 > T_0/T$, we have

$$c_1 T > 2\sqrt{\kappa_1} \geq 2 \Big(\sum_{j,k=1}^n h^{jk} \psi_{x_j} \psi_{x_k} \Big)^{\frac{1}{2}} \geq 2\sqrt{h_0} |\nabla \psi|. \tag{4.36}$$

This implies that

$$M(0, x) \geq \left[\frac{1}{2} \lambda c_1 T \min\{1, h_0\} (v_t^2 + |\nabla v|^2) + \frac{3}{4} \lambda^3 c_1^3 T^3 v^2 + O(\lambda^2) v^2 - v_t^2 \right]\Big|_{t=0}. \tag{4.37}$$

Further (recall that $v = \theta z$),

$$2s_0\lambda z_t\eta_1\cdot\nabla z\Big|_{t=0}$$
$$\geq -\sqrt{n}s_0\lambda e^{2\lambda\phi}\big(z_t^2+|\nabla z|^2\big)\Big|_{t=0} \tag{4.38}$$
$$\geq -2\sqrt{n}s_0\lambda\big(v_t^2+\lambda^2 c_1^2 T^2 v^2+|\nabla v|^2+\lambda^2|\nabla\psi|^2 v^2\big)\Big|_{t=0}.$$

On the other hand,

$$-\frac{1}{h_0}\Big[12\sqrt{n}\lambda^2 s_0\Big(|\nabla\psi|\sum_{j,k=1}^n |h^{jk}|_{C(\overline{\Omega})}+1\Big)\Big]\eta_2 z z_t$$

$$=-\frac{1}{h_0}\Big[12\sqrt{n}\lambda^2 s_0\Big(|\nabla\psi|\sum_{j,k=1}^n |h^{jk}|_{C(\overline{\Omega})}+1\Big)\Big]\rho_2^2\Big(vv_t-\lambda c_1 T v^2-\frac{v_t^2}{\lambda c_1 T}+\frac{v_t^2}{\lambda c_1 T}\Big)\Big|_{t=0}$$

$$\geq -\frac{1}{h_0 c_1 T}\lambda\Big[12\sqrt{n}s_0\Big(|\nabla\psi|\sum_{j,k=1}^n |h^{jk}|_{C(\overline{\Omega})}+1\Big)\Big]v_t^2\Big|_{t=0}. \tag{4.39}$$

It follows from (4.37)–(4.39) that

$$\overline{M}(0,x)\geq\Big(\lambda F_1 v_t^2+\lambda F_2|\nabla v|^2+\lambda^3 F_3 v^2+O(\lambda^2)v^2+O(1)v_t^2\Big)\Big|_{t=0}, \tag{4.40}$$

where

$$\begin{cases} F_1=\dfrac{1}{2}c_1 T\min\{1,h_0\}-2\sqrt{n}s_0-\dfrac{1}{h_0 c_1 T}\Big[12\sqrt{n}s_0\Big(|\nabla\psi|\displaystyle\sum_{j,k=1}^n |h^{jk}|_{C(\overline{\Omega})}+1\Big)\Big], \\[4mm] F_2=\dfrac{1}{2}c_1 T\min\{1,h_0\}-2\sqrt{n}s_0,\quad F_3=\dfrac{3}{4}c_1^3 T^3-2\sqrt{n}s_0\big(c_1^2 T^2+|\nabla\psi|^2\big). \end{cases}$$

By (4.28) and (4.36), for any $T>T_0$, it holds that $c_1 T>1$,

$$F_1\geq\frac{1}{2}c_1 T\min\{1,h_0\}-2\sqrt{n}s_0-\frac{6\sqrt{n}s_0}{h_0^{3/2}}\sum_{j,k=1}^n |h^{jk}|_{C(\overline{\Omega})}-\frac{12\sqrt{n}s_0}{h_0}>0,$$

and therefore, $F_2>0$. Moreover,

$$F_3\geq\frac{3}{4}c_1^3 T^3-2\sqrt{n}s_0 c_1^2 T^2\Big(1+\frac{1}{4h_0}\Big)$$
$$=\frac{3}{4}c_1^2 T^2\Big[c_1 T-\frac{8}{3}\sqrt{n}s_0\Big(1+\frac{1}{4h_0}\Big)\Big]>0,$$

where we have used the following fact:

$$\begin{aligned}
c_1 T &> \frac{4\sqrt{n}}{\min\{1, h_0\}} s_0 = \frac{8\sqrt{n}}{3\min\{1, h_0\}} s_0 + \frac{4\sqrt{n}}{3\min\{1, h_0\}} s_0 \\
&\geq \frac{8\sqrt{n}}{3} s_0 + \frac{4\sqrt{n}}{3h_0} s_0 = \frac{8}{3}\sqrt{n} s_0 \left(1 + \frac{1}{2h_0}\right).
\end{aligned}$$

Finally, by (4.28), one can find constants $C > 0$ and $\lambda_2 > 0$ such that for any $\lambda \geq \lambda_2$,

$$\overline{M}(0, \cdot) \geq 0. \tag{4.41}$$

Meanwhile, noting that $\ell(T, x) = -\ell(0, x)$, we may find a constant $\lambda_3 > 0$ such that for any $\lambda \geq \lambda_3$,

$$\overline{M}(T, \cdot) \leq 0. \tag{4.42}$$

Combining (4.41) and (4.42) with (4.35), and noting that $v = \theta z$, for any $\lambda \geq \lambda_0 \overset{\triangle}{=} \max\{\lambda_1, \lambda_2, \lambda_3\}$, we obtain the desired estimate (4.30). This completes the proof of Theorem 4.2.

Remark 4.4 Notice that Theorem 4.2 is an improvement of [12, Theorem 5.1]. Indeed, in [12, Theorem 5.1], the global Carleman estimate was established under the additional condition that $z(0, \cdot) = z(T, \cdot) = 0$ in Ω. However, this condition seems too restrictive to be fulfilled in applications (e.g., the Eq. (7.5) in [12]). Therefore, it is necessary to prove (4.30) without such a restriction.

4.1.2 Carleman Estimate in L^2-Norm

To obtain the desired Carleman estimate in L^2-Norm, we first introduce an auxiliary optimal control problem.

Throughout this section, we fix ϕ as in (4.10), a parameter $\lambda > 0$ and a function $z \in C([0, T]; L^2(\Omega))$ satisfying $z(0, \cdot) = z(T, \cdot) = 0$ in Ω. For any $K > 1$, we choose $\rho \equiv \rho^K(x) \in C^2(\overline{\Omega})$ with $\min_{x \in \Omega} \rho(x) = 1$ such that (recall (4.9) for ω)

$$\rho(x) = \begin{cases} 1, & \text{for } x \in \omega, \\ K, & \text{for dist}(x, \omega) \geq \frac{1}{\ln K}. \end{cases} \tag{4.43}$$

Let $\tau = \frac{T}{m}$ for a fixed integer $m \geq 3$. Define

$$z_m^j \equiv z_m^j(x) = z(j\tau, x), \qquad \phi_m^j \equiv \phi_m^j(x) = \phi(j\tau, x), \quad j = 0, 1, \cdots, m. \tag{4.44}$$

Let $\{(w_m^j, r_{1m}^j, r_{2m}^j, r_m^j)\}_{j=0}^m \in (H_0^1(\Omega) \times (L^2(\Omega))^3)^{m+1}$ solving the following system

$$
\begin{cases}
\dfrac{w_m^{j+1} - 2w_m^j + w_m^{j-1}}{\tau^2} - \displaystyle\sum_{j_1,j_2=1}^{n} \partial_{x_{j_2}}(h^{j_1 j_2} \partial_{x_{j_1}} w_m^j) \\
\quad = \dfrac{r_{1m}^{j+1} - r_{1m}^j}{\tau} + r_{2m}^j + \lambda z_m^j e^{2\lambda \phi_m^j} + r_m^j, \quad (1 \le j \le m-1), \ \text{in } \Omega, \\
w_m^j = 0, \quad (0 \le j \le m), \quad\quad\quad\quad\quad\quad\quad\quad \text{on } \Gamma, \\
w_m^0 = w_m^m = r_{2m}^0 = r_{2m}^m = r_m^0 = r_m^m = 0, \quad r_{1m}^0 = r_{1m}^1, \quad \text{in } \Omega.
\end{cases}
\tag{4.45}
$$

Note that, here neither r_{1m}^0 nor r_{1m}^m is assumed to be zero. Instead we assume that $r_{1m}^0 = r_{1m}^1$. In the system (4.45), $(r_{1m}^j, r_{2m}^j, r_m^j) \in (L^2(\Omega))^3$ $(j = 0, 1, \cdots, m)$ can be regarded as controls. The set of *admissible sequences* for (4.45) is defined as

$$
\mathscr{A}_{ad} \stackrel{\triangle}{=} \Big\{ \{(w_m^j, r_{1m}^j, r_{2m}^j, r_m^j)\}_{j=0}^m \in (H_0^1(\Omega) \times (L^2(\Omega))^3)^{m+1} : \\
\{(w_m^j, r_{1m}^j, r_{2m}^j, r_m^j)\}_{j=0}^m \text{ fulfills } (4.45) \Big\}.
$$

Clearly, $\mathscr{A}_{ad} \ne \emptyset$ since $\{(0, 0, 0, -\lambda z_m^j e^{2\lambda \phi_m^j})\}_{j=0}^m \in \mathscr{A}_{ad}$.

Next, we introduce the cost functional:

$$
\begin{aligned}
\mathscr{J} (&\{(w_m^j, r_{1m}^j, r_{2m}^j, r_m^j)\}_{j=0}^m) \\
&= \frac{\tau}{2} \int_\Omega \rho \frac{|r_{1m}^m|^2}{\lambda^2} e^{-2\lambda \phi_m^m} dx + \frac{\tau}{2} \sum_{\ell=1}^{m-1} \Big[\int_\Omega |w_m^j|^2 e^{-2\lambda \phi_m^j} dx \\
&\quad + \int_\Omega \rho \Big(\frac{|r_{1m}^j|^2}{\lambda^2} + \frac{|r_{2m}^j|^2}{\lambda^4} \Big) e^{-2\lambda \phi_m^j} dx + K \int_\Omega |r_m^j|^2 dx \Big]
\end{aligned}
\tag{4.46}
$$

and consider the following *optimal control problem*:

Problem (OP): Find a $\{(\hat{w}_m^j, \hat{r}_{1m}^j, \hat{r}_{2m}^j, \hat{r}_m^j)\}_{j=0}^m \in \mathscr{A}_{ad}$, such that

$$
\begin{aligned}
\mathscr{J} (&\{(\hat{w}_m^j, \hat{r}_{1m}^j, \hat{r}_{2m}^j, \hat{r}_m^j)\}_{j=0}^m) \\
&= \inf_{\{(w_m^j, r_{1m}^j, r_{2m}^j, r_m^j)\}_{j=0}^m \in \mathscr{A}_{ad}} \mathscr{J} (\{(w_m^j, r_{1m}^j, r_{2m}^j, r_m^j)\}_{j=0}^m).
\end{aligned}
\tag{4.47}
$$

For any $\{(w_m^j, r_{1m}^j, r_{2m}^j, r_m^j)\}_{j=0}^m \in \mathscr{A}_{ad}$, by the standard regularity results for elliptic equations, one has that $w_m^j \in H^2(\Omega) \cap H_0^1(\Omega)$ for $j = 0, \cdots, m$. We need the following known technical result (See [12, Proposition 6.1] and [14, pp. 190–199] for its proof):

Proposition 4.1 *For any $K > 1$ and $m \ge 3$, Problem (OP) admits a unique solution $\{(\hat{w}_m^j, \hat{r}_{1m}^j, \hat{r}_{2m}^j, \hat{r}_m^j)\}_{j=0}^m \in \mathscr{A}_{ad}$ (which depends on K). Furthermore, for*

$$
p_m^j \equiv p_m^j(x) \stackrel{\triangle}{=} K \hat{r}_m^j(x), \quad 0 \le j \le m,
\tag{4.48}
$$

one has

$$\hat{w}_m^0 = \hat{w}_m^m = p_m^0 = p_m^m = 0 \; in \; \Omega,$$
$$\hat{w}_m^j, p_m^j \in H^2(\Omega) \cap H_0^1(\Omega) \; for \; 1 \le j \le m-1, \tag{4.49}$$

and the following optimality conditions hold:

$$\begin{cases} \dfrac{p_m^j - p_m^{j-1}}{\tau} + \rho \dfrac{\hat{r}_{1m}^j}{\lambda^2} e^{-2\lambda\phi_m^j} = 0 & in\, \Omega, \\[4mm] p_m^j - \rho \dfrac{\hat{r}_{2m}^j}{\lambda^4} e^{-2\lambda\phi_m^j} = 0 & in\; \Omega, \end{cases} \qquad 1 \le j \le m, \tag{4.50}$$

$$\begin{cases} \dfrac{p_m^{j+1} - 2p_m^j + p_m^{j-1}}{\tau^2} - \displaystyle\sum_{j_1,j_2=1}^n \partial_{x_{j_2}} (h^{j_1 j_2} \partial_{x_{j_1}} p_m^j) + \hat{w}_m^j e^{-2\lambda\phi_m^j} = 0 \; in\; \Omega, \\[4mm] p_m^j = 0 \hfill on\; \Gamma, \\[3mm] \hspace{4cm} 1 \le j \le m-1. \end{cases} \tag{4.51}$$

Moreover, there is a constant $C = C(K, \lambda) > 0$, independent of m, such that

$$\tau \sum_{j=1}^{m-1} \int_\Omega \left(|\hat{w}_m^j|^2 + |\hat{r}_{1m}^j|^2 + |\hat{r}_{2m}^j|^2 + K|\hat{r}_m^j|^2 \right) dx + \tau \int_\Omega |\hat{r}_{1m}^m|^2 dx \le C, \tag{4.52}$$

and

$$\tau \sum_{j=0}^{m-1} \int_\Omega \left[\frac{(\hat{w}_m^{j+1} - \hat{w}_m^j)^2}{\tau^2} + \frac{(\hat{r}_{1m}^{j+1} - \hat{r}_{1m}^j)^2}{\tau^2} + \frac{(\hat{r}_{2m}^{j+1} - \hat{r}_{2m}^j)^2}{\tau^2} \right. $$
$$\left. + K \frac{(\hat{r}_m^{j+1} - \hat{r}_m^j)^2}{\tau^2} \right] dx \le C. \tag{4.53}$$

Now, let us consider the following hyperbolic equation:

$$\begin{cases} z_{tt} - \displaystyle\sum_{j,k=1}^n (h^{jk} z_{x_j})_{x_k} = f \; in \; Q, \\[3mm] z = 0 \hfill on \; \Sigma, \end{cases} \tag{4.54}$$

where $f \in L^1(0, T; H^{-1}(\Omega))$. We call $z \in L^2(Q)$ a *weak solution* to (4.54) if

$$\left\langle z, \eta_{tt} - \sum_{j,k=1}^n (h^{jk} \eta_{x_j})_{x_k} \right\rangle_{L^2(Q)} = \int_0^T \left\langle f(t, \cdot), \eta(t, \cdot) \right\rangle_{H^{-1}(\Omega), H_0^1(\Omega)} dt, \tag{4.55}$$
$$\forall \, \eta \in H_0^2((0, T); H^2(\Omega) \cap H_0^1(\Omega)).$$

Note that, there are no initial conditions in (4.54). One can prove the following regularity result for weak solutions to (4.54) (See [12, Lemma 3.1]).

Lemma 4.1 *Let $0 < t_1 < t_2 < T$, and $g \in L^2((t_1, t_2) \times \Omega)$ be given. Assume that $z \in L^2(Q)$ is a weak solution to (4.54), and $z = g$ in $(t_1, t_2) \times \Omega$. Then $z \in C([0, T]; L^2(\Omega)) \cap C^1([0, T]; H^{-1}(\Omega))$, and there exists a constant $C > 0$, depending only on T, t_1, t_2, Ω, and $(h^{jk})_{1 \le j,k \le n}$, such that*

$$|z|_{C([0,T];L^2(\Omega)) \cap C^1([0,T];H^{-1}(\Omega))} \le C\big(|f|_{L^1(0,T;H^{-1}(\Omega))} + |g|_{L^2((t_1,t_2)\times\Omega)}\big). \quad (4.56)$$

Our Carleman estimate for hyperbolic operators in L^2-norm is as follows:

Theorem 4.3 *Let $a \in L^\infty(0, T; L^p(\Omega))$ for some $p \in [n, \infty]$. Suppose that Condition 4.2 holds, and ω and T_1 are given in (4.9) and (4.28), respectively. Then there exists $\lambda_0^* > 0$ such that for all $T > T_1$, $\lambda \ge \lambda_0^*$, and every $z \in C([0, T]; L^2(\Omega))$ satisfying $z(0, \cdot) = z(T, \cdot) = 0$ in Ω and $z_{tt} - \sum_{j,k=1}^n (h^{jk} z_{x_j})_{x_k} \in H^{-1}(Q)$, it holds*

$$\lambda|\theta z|_{L^2(Q)}^2 \le C\bigg\{\bigg|\theta\bigg[z_{tt} - \sum_{j,k=1}^n (h^{jk} z_{x_j})_{x_k} - az\bigg]\bigg|_{H^{-1}(Q)}^2 + \lambda^2|\theta z|_{((0,T)\times\omega)}^2$$

$$+ \frac{1}{\lambda^{2(1-n/p)}}|\theta az|_{L^2(0,T;H^{-n/p}(\Omega))}^2\bigg\}, \quad (4.57)$$

where ϕ is given by (4.10).

Proof The main idea is to apply (4.55) to some special η with $\eta_{tt} - \sum_{j,k=1}^n \partial_{x_k}(h^{jk}\eta_{x_j})$
$= \cdots + \lambda z e^{2\lambda\phi}$, which yields the desired term $\lambda \int_Q \theta^2 z^2 dxdt$ and reduces the estimate to that for $|\eta|_{H_0^1(Q)}$. The proof is divided into in six steps.

  Step 1. First of all, recall the functions $\{(\hat{w}_m^j, \hat{r}_{1m}^j, \hat{r}_{2m}^j, \hat{r}_m^j)\}_{j=0}^m$ in Proposition 4.1. Put

$$\begin{cases} \tilde{w}^m(t, x) = \dfrac{1}{\tau}\sum_{j=0}^{m-1}\Big\{(t - j\tau)\hat{w}_m^{j+1}(x) - [t - (j+1)\tau]\hat{w}_m^j(x)\Big\}\chi_{(j\tau,(j+1)\tau]}(t), \\[2mm] \tilde{r}_1^m(t, x) = \hat{r}_{1m}^0(x)\chi_{\{0\}}(t) \\[1mm] \qquad\qquad + \dfrac{1}{\tau}\sum_{j=0}^{m-1}\Big\{(t - j\tau)\hat{r}_{1m}^{j+1}(x) - [t - (j+1)\tau]\hat{r}_{1m}^j(x)\Big\}\chi_{(j\tau,(j+1)\tau]}(t), \\[2mm] \tilde{r}_2^m(t, x) = \dfrac{1}{\tau}\sum_{j=0}^{m-1}\Big\{(t - j\tau)\hat{r}_{2m}^{j+1}(x) - [t - (j+1)\tau]\hat{r}_{2m}^j(x)\Big\}\chi_{(j\tau,(j+1)\tau]}(t), \\[2mm] \tilde{r}^m(t, x) = \dfrac{1}{\tau}\sum_{j=0}^{m-1}\Big\{(t - j\tau)\hat{r}_m^{j+1}(x) - [t - (j+1)\tau]\hat{r}_m^j(x)\Big\}\chi_{(j\tau,(j+1)\tau]}(t). \end{cases}$$

By (4.52) and (4.53), one can find a subsequence of $\{(\tilde{w}^m, \tilde{r}_1^m, \tilde{r}_2^m, \tilde{r}^m)\}_{m=1}^\infty$, which converges weakly to some $(\tilde{w}, \tilde{r}_1, \tilde{r}_2, \tilde{r}) \in (H^1(0, T; L^2(\Omega)))^4$, as $m \to \infty$.

For any constant $K > 1$, put $\tilde{p} \overset{\triangle}{=} K\tilde{r}$. In what follows, we shall choose K to be sufficiently large (see (4.71)). By (4.45), (4.50)–(4.53) and using Lemma 4.1, we see that $\tilde{w}$, $\tilde{p} \in C([0, T]; H_0^1(\Omega)) \cap C^1([0, T]; L^2(\Omega))$, and

$$
\begin{cases}
\tilde{w}_{tt} - \displaystyle\sum_{j,k=1}^{n} (h^{jk}\tilde{w}_{x_j})_{x_k} = \tilde{r}_{1,t} + \tilde{r}_2 + \lambda\theta^2 z + \tilde{r} & \text{in } Q, \\[2mm]
\tilde{p}_{tt} - \displaystyle\sum_{j,k=1}^{n} (h^{jk}\tilde{p}_{x_j})_{x_k} + \theta^{-2}\tilde{w} = 0 & \text{in } Q, \\[2mm]
\tilde{p} = \tilde{w} = 0 & \text{on } \Sigma, \\[1mm]
\tilde{p}(0) = \tilde{p}(T) = \tilde{w}(0) = \tilde{w}(T) = 0 & \text{in } \Omega, \\[2mm]
\tilde{p}_t + \rho\theta^{-2}\dfrac{\tilde{r}_1}{\lambda^2} = 0 & \text{in } Q, \\[3mm]
\tilde{p} - \rho\theta^{-2}\dfrac{\tilde{r}_2}{\lambda^4} = 0 & \text{in } Q.
\end{cases}
\tag{4.58}
$$

Step 2. Applying Theorem 4.2 to $\tilde{p}$ in (4.58), one gets that

$$
\begin{aligned}
&\lambda \int_Q \theta^2\big(\lambda^2\tilde{p}^2 + \tilde{p}_t^2 + |\nabla\tilde{p}|^2\big)dxdt \\
&\leq C\Big[\int_Q \theta^{-2}\tilde{w}^2 dxdt + \lambda^2 \int_0^T \int_\omega \theta^2(\lambda^2\tilde{p}^2 + \tilde{p}_t^2)dxdt\Big] \\
&\leq C\Big[\int_Q \theta^{-2}\tilde{w}^2 dxdt + \int_0^T \int_\omega \theta^{-2}\Big(\frac{\tilde{r}_1^2}{\lambda^2} + \frac{\tilde{r}_2^2}{\lambda^4}\Big)dxdt\Big].
\end{aligned}
\tag{4.59}
$$

Here and henceforth, C is a constant, independent of K and λ.

By (4.58) again, one finds that $\tilde{p}_t$ solves

$$
\begin{cases}
\tilde{p}_{ttt} - \displaystyle\sum_{j,k=1}^{n} (h^{jk}\tilde{p}_{tx_j})_{x_k} + (\theta^{-2}\tilde{w})_t = 0 & \text{in } Q, \\[2mm]
\tilde{p}_t = 0 & \text{on } \Sigma, \\[2mm]
\tilde{p}_{tt} + \dfrac{\rho}{\lambda}\theta^{-2}\Big(\dfrac{\tilde{r}_{1,t}}{\lambda} - 2\phi_t\tilde{r}_1\Big) = 0 & \text{in } Q, \\[3mm]
\tilde{p}_t - \dfrac{\rho}{\lambda^2}\theta^{-2}\Big(\dfrac{\tilde{r}_{2,t}}{\lambda^2} - \dfrac{2}{\lambda}\phi_t\tilde{r}_2\Big) = 0 & \text{in } Q.
\end{cases}
\tag{4.60}
$$

Applying Theorem 4.2 to $\tilde{p}_t$ and noting (4.60), we obtain that

$$
\begin{aligned}
&\lambda \int_Q \theta^2\big(\lambda^2\tilde{p}_t^2 + \tilde{p}_{tt}^2 + |\nabla\tilde{p}_t|^2\big)dxdt \\
&\leq C\Big[|\theta(\theta^{-2}\tilde{w})_t|^2_{L^2(Q)} + \lambda^2 \int_0^T \int_\omega \theta^2(\lambda^2\tilde{p}_t^2 + \tilde{p}_{tt}^2)dxdt\Big] \\
&\leq C\Big[\int_Q \theta^{-2}(\tilde{w}_t^2 + \lambda^2\tilde{w}^2)dxdt + \int_0^T \int_\omega \theta^{-2}\Big(\frac{\tilde{r}_{1,t}^2}{\lambda^2} + \frac{\tilde{r}_{2,t}^2}{\lambda^4} + \tilde{r}_1^2 + \frac{\tilde{r}_2^2}{\lambda^2}\Big)dxdt\Big].
\end{aligned}
\tag{4.61}
$$

Step 3. From (4.58), and noting that

$$-\int_Q (\tilde{r}_{1,t} + \tilde{r}_2)\tilde{p}\,dxdt = \int_Q (\tilde{r}_1\tilde{p}_t - \tilde{r}_2\tilde{p})\,dxdt = -\int_Q \rho\theta^{-2}\Big(\frac{\tilde{r}_1^2}{\lambda^2} + \frac{\tilde{r}_2^2}{\lambda^4}\Big)dxdt,$$

and recalling $\tilde{p} = K\tilde{r}$, we get

$$0 = \Big\langle \tilde{w}_{tt} - \sum_{j,k=1}^n \big(h^{jk}\tilde{w}_{x_j}\big)_{x_k} - \tilde{r}_{1,t} - \tilde{r}_2 - \lambda\theta^2 z - \tilde{r}, \tilde{p}\Big\rangle_{L^2(Q)}$$

$$= -\int_Q \theta^{-2}\tilde{w}^2\,dxdt - \int_Q \rho\theta^{-2}\Big(\frac{\tilde{r}_1^2}{\lambda^2} + \frac{\tilde{r}_2^2}{\lambda^4}\Big)dxdt - \lambda\int_Q \theta^2 z\tilde{p}\,dxdt - K\int_Q \tilde{r}^2\,dxdt.$$

Hence

$$\int_Q \theta^{-2}\tilde{w}^2\,dxdt + \int_Q \rho\theta^{-2}\Big(\frac{\tilde{r}_1^2}{\lambda^2} + \frac{\tilde{r}_2^2}{\lambda^4}\Big)dxdt + K\int_Q \tilde{r}^2\,dxdt = -\lambda\int_Q \theta^2 z\tilde{p}\,dxdt.$$

This, together with (4.59), implies that

$$\int_Q \theta^{-2}\tilde{w}^2\,dxdt + \int_Q \rho\theta^{-2}\Big(\frac{\tilde{r}_1^2}{\lambda^2} + \frac{\tilde{r}_2^2}{\lambda^4}\Big)dxdt + K\int_Q \tilde{r}^2\,dxdt \le \frac{C}{\lambda}\int_Q \theta^2 z^2\,dxdt.$$
$$(4.62)$$

Step 4. Using (4.58) and (4.60) again, and noting $\tilde{p}_{tt}(0,\cdot) = \tilde{p}_{tt}(T,\cdot) = 0$ in Ω, we get that

$$0 = \Big\langle \tilde{w}_{tt} - \sum_{j,k=1}^n \partial_{x_k}(h^{jk}\tilde{w}_{x_j}) - \tilde{r}_{1,t} - \tilde{r}_2 - \lambda\theta^2 z - \tilde{r}, \tilde{p}_{tt}\Big\rangle_{L^2(Q)} \qquad (4.63)$$

$$= -\int_Q \tilde{w}\big(\theta^{-2}\tilde{w}\big)_{tt}\,dxdt - \int_Q (\tilde{r}_{1,t} + \tilde{r}_2)\tilde{p}_{tt}\,dxdt - \lambda\int_Q \theta^2 z\tilde{p}_{tt}\,dxdt - \int_Q \tilde{r}\tilde{p}_{tt}\,dxdt.$$

Clearly,

$$-\int_Q \tilde{w}\big(\theta^{-2}\tilde{w}\big)_{tt}\,dxdt = \int_Q \Big[\theta^{-2}\tilde{w}_t^2 - (\theta^{-2})_{tt}\frac{\tilde{w}^2}{2}\Big]dxdt$$

$$= \int_Q \theta^{-2}\big(\tilde{w}_t^2 + \lambda\phi_{tt}\tilde{w}^2 - 2\lambda^2\phi_t^2\tilde{w}^2\big)dxdt.$$
$$(4.64)$$

Further, in view of the third and fourth equalities in (4.60), one has

$$-\int_Q (\tilde{r}_{1,t} + \tilde{r}_2)\tilde{p}_{tt}\,dxdt$$

$$= -\int_Q (\tilde{r}_{1,t}\tilde{p}_{tt} - \tilde{r}_{2,t}\tilde{p}_t)\,dxdt$$

$$= \int_Q \theta^{-2}\tilde{r}_{1,t}\frac{\rho}{\lambda}\Big(\frac{\tilde{r}_{1,t}}{\lambda} - 2\phi_t\tilde{r}_1\Big)dxdt + \int_Q \theta^{-2}\tilde{r}_{2,t}\frac{\rho}{\lambda^2}\Big(\frac{\tilde{r}_{2,t}}{\lambda^2} - \frac{2}{\lambda}\phi_t\tilde{r}_2\Big)dxdt \qquad (4.65)$$

$$= \int_Q \rho\theta^{-2}\Big(\frac{\tilde{r}_{1,t}^2}{\lambda^2} + \frac{\tilde{r}_{2,t}^2}{\lambda^4} - \frac{2}{\lambda}\phi_t\tilde{r}_1\tilde{r}_{1,t} - \frac{2}{\lambda^3}\phi_t\tilde{r}_2\tilde{r}_{2,t}\Big)dxdt.$$

Moreover, by $\tilde{p} = K\tilde{r}$ and integration by parts, one gets that

$$-\int_Q \tilde{r}\tilde{p}_{tt}\,dxdt = K\int_Q \tilde{r}_t^2\,dxdt. \qquad (4.66)$$

Combining (4.63)–(4.66), we end up with

$$\int_Q \rho\theta^{-2}\Big(\frac{\tilde{r}_{1,t}^2}{\lambda^2} + \frac{\tilde{r}_{2,t}^2}{\lambda^4} - \frac{2}{\lambda}\phi_t\tilde{r}_1\tilde{r}_{1,t} - \frac{2}{\lambda^3}\phi_t\tilde{r}_2\tilde{r}_{2,t}\Big)dxdt + K\int_Q \tilde{r}_t^2\,dxdt$$
$$+ \int_Q \theta^{-2}(\tilde{w}_t^2 + \lambda\phi_{tt}\tilde{w}^2 - 2\lambda^2\phi_t^2\tilde{w}^2)dxdt = \lambda\int_Q \theta^2 z\tilde{p}_{tt}\,dxdt. \qquad (4.67)$$

Now, by $(4.67) + C\lambda^2 \times (4.62)$ (with a sufficiently large $C > 0$), using Cauchy-Schwartz inequality and noting (4.61), we obtain that

$$\int_Q \theta^{-2}(\tilde{w}_t^2 + \lambda^2\tilde{w}^2)dxdt + \int_Q \rho\theta^{-2}\Big(\frac{\tilde{r}_{1,t}^2}{\lambda^2} + \frac{\tilde{r}_{2,t}^2}{\lambda^4} + \tilde{r}_1^2 + \frac{\tilde{r}_2^2}{\lambda^2}\Big)dxdt \leq C\lambda\int_Q \theta^2 z^2\,dxdt. \qquad (4.68)$$

Step 5. It follows from (4.58) that

$$\big\langle \tilde{r}_{1,t} + \tilde{r}_2 + \lambda\theta^2 z + \tilde{r},\ \theta^{-2}\tilde{w}\big\rangle_{L^2(Q)}$$

$$= \Big\langle \tilde{w}_{tt} - \sum_{j,k=1}^n \big(h^{jk}\tilde{w}_{x_j}\big)_{x_k},\ \theta^{-2}\tilde{w}\Big\rangle_{L^2(Q)}$$

$$= -\int_Q \tilde{w}_t\big(\theta^{-2}\tilde{w}\big)_t\,dxdt + \sum_{j,k=1}^n \int_Q h^{jk}\tilde{w}_{x_j}\big(\theta^{-2}\tilde{w}\big)_{x_k}\,dxdt \qquad (4.69)$$

$$= -\int_Q \theta^{-2}(\tilde{w}_t^2 + \lambda\phi_{tt}\tilde{w}^2 - 2\lambda^2\phi_t^2\tilde{w}^2)dxdt + \sum_{j,k=1}^n \int_Q \theta^{-2}h^{jk}\tilde{w}_{x_j}\tilde{w}_{x_k}\,dxdt$$

$$-2\lambda\sum_{j,k=1}^n \int_Q \theta^{-2}h^{jk}\tilde{w}_{x_j}\tilde{w}\phi_{x_k}\,dxdt.$$

This, together with (4.2), yields (recall $\lambda \geq \lambda_0 > 1$)

$$\int_Q \theta^{-2} |\nabla \tilde{w}|^2 dx dt$$

$$\leq C \int_Q \left[\theta^{-2}\big((|\tilde{r}_{1,t} + \tilde{r}_2 + \tilde{r}|)|\tilde{w}|\big) + \lambda |z\tilde{w}| + \theta^{-2}(\tilde{w}_t^2 + \lambda^2 \tilde{w}^2)\right] dx dt \qquad (4.70)$$

$$\leq C \int_Q \left[\theta^2 z^2 + \theta^{-2}\Big(\frac{\tilde{r}_{1,t}^2}{\lambda^2} + \frac{\tilde{r}_2^2}{\lambda^2} + \tilde{r}^2 + \tilde{z}_t^2 + \lambda^2 \tilde{w}^2\Big)\right] dx dt.$$

Combining (4.62), (4.68) and (4.70), choosing the constant K in (4.62) be such that

$$K \geq C e^{2\lambda \max_{(t,x) \in \overline{Q}} |\phi|} \qquad (4.71)$$

(to absorb the term $C \int_Q \tilde{r}^2 \theta^{-2} dx dt$ in the right hand side of (4.70)), and noting that $\rho(x) \geq 1$ in Ω, we deduce that

$$\int_Q \theta^{-2} (|\nabla \tilde{w}|^2 + \tilde{w}_t^2 + \lambda^2 \tilde{w}^2) dx dt + \int_Q \theta^{-2} \rho \Big(\frac{\tilde{r}_{1,t}^2}{\lambda^2} + \frac{\tilde{r}_{2,t}^2}{\lambda^4} + \tilde{r}_1^2 + \frac{\tilde{r}_2^2}{\lambda^2}\Big) dx dt$$

$$\leq C\lambda \int_Q \theta^2 z^2 dx dt. \qquad (4.72)$$

Step 6. Recall that $(\tilde{w}, \tilde{r}_1, \tilde{r}_2, \tilde{r})$ depends on K. Now we denote it by $(\tilde{w}^K, \tilde{r}_1^K, \tilde{r}_2^K, \tilde{r}^K)$ to emphasize this dependence. Fix λ and let $K \to \infty$. Since $\rho(x) \equiv \rho^K(x) \to \infty$ for any $x \notin \omega$, as $K \to \infty$, it follows from (4.62) and (4.72) that there exists a subsequence of $(\tilde{w}^K, \tilde{r}_1^K, \tilde{r}_2^K, \tilde{r}^K)$ which converges weakly to some $(\check{w}, \check{r}_1, \check{r}_2, 0)$ in $H_0^1(Q) \times (H^1(0, T; L^2(\Omega)))^2 \times L^2(Q)$, with $\operatorname{supp} \check{r}_j \subset \overline{(0, T) \times \omega}$ $(j = 1, 2)$. By (4.58), we deduce that $(\check{w}, \check{r}_1, \check{r}_2)$ satisfies

$$\begin{cases} \check{w}_{tt} - \displaystyle\sum_{j,k=1}^{n} (h^{jk} \check{w}_{x_j})_{x_k} = \check{r}_{1,t} + \check{r}_2 + \lambda \theta^2 z & \text{in } Q, \\ \check{w} = 0 & \text{on } \partial Q. \end{cases} \qquad (4.73)$$

Using (4.72) again, we find that

$$|\theta^{-1} \check{w}|^2_{H_0^1(Q)} + \frac{1}{\lambda^2} \int_0^T \int_\omega \theta^{-2}(\check{r}_{1,t}^2 + \check{r}_2^2) dx dt \leq C\lambda \int_Q \theta^2 z^2 dx dt. \qquad (4.74)$$

Now, by (4.55) with η replaced by the above $\check{w}$, one gets that

$$\Big\langle \check{w}, \check{r}_{1,t} + \check{r}_2 + \lambda z \theta^2 \Big\rangle_{L^2(Q)} = \Big\langle z_{tt} - \sum_{j,k=1}^{n} (h^{jk} z_{x_j})_{x_k}, \check{w} \Big\rangle_{H^{-1}(Q), H_0^1(Q)}.$$

Hence, noting $\operatorname{supp} \check{r}_j \subset \overline{(0, T) \times \omega}$ $(j = 1, 2)$, it holds that

$$\lambda \int_Q \theta^2 z^2 dx dt$$

$$= \left\langle z_{tt} - \sum_{j,k=1}^{n} (h^{jk} z_{x_j})_{x_k} - az, \check{w} \right\rangle_{H^{-1}(Q), H_0^1(Q)} + \langle az, \check{w} \rangle_{L^2(Q)} - \langle z, \check{r}_{1,t} + \check{r}_2 \rangle_{L^2((0,T)\times\omega)}$$

$$\leq \left| \theta \left[z_{tt} - \sum_{j,k=1}^{n} (h^{jk} z_{x_j})_{x_k} - az \right] \right|_{H^{-1}(Q)} |\theta^{-1} \check{w}|_{H_0^1(Q)} \tag{4.75}$$

$$+ |\theta az|_{L^2(0,T;H^{-n/p}(\Omega))} |\theta^{-1} \check{w}|_{L^2(0,T;H_0^{n/p}(\Omega))}$$

$$+ |\theta z|_{L^2((0,T)\times\omega)} |\theta^{-1}(\check{r}_{1,t} + \check{r}_2)|_{L^2((0,T)\times\omega)}$$

$$\leq C\sqrt{\mathscr{Q}} \Big[|\theta^{-1} \check{w}|^2_{H_0^1(Q)} + \lambda^{2(1-n/p)} |\theta^{-1} \check{w}|^2_{L^2(0,T;H_0^{n/p}(\Omega))}$$

$$+ \frac{1}{\lambda^2} |\theta^{-1}(\check{r}_{1,t} + \check{r}_2)|^2_{L^2((0,T)\times\omega)} \Big]^{1/2},$$

where

$$\mathscr{Q} \stackrel{\triangle}{=} \left| \theta \left[z_{tt} - \sum_{j,k=1}^{n} \partial_{x_k} (h^{jk} z_{x_j}) - az \right] \right|^2_{H^{-1}(Q)}$$

$$+ \frac{1}{\lambda^{2(1-n/p)}} |\theta az|^2_{L^2(0,T;H^{-n/p}(\Omega))} + \lambda^2 |\theta z|^2_{L^2((0,T)\times\omega)}$$

is the right hand side of (4.57).

It follows from (3.56) and using Young's inequality that

$$\lambda^{2(1-n/p)} |\theta^{-1} \check{w}|^2_{L^2(0,T;H_0^{n/p}(\Omega))}$$

$$\leq C\lambda^{2(1-n/p)} |\theta^{-1} \check{w}|^{2n/p}_{L^2(0,T;H_0^1(\Omega))} |\theta^{-1} \check{w}|^{2(1-n/p)}_{L^2(Q)} \tag{4.76}$$

$$\leq C \Big(|\theta^{-1} \check{w}|^2_{L^2(0,T;H_0^1(\Omega))} + \lambda^2 |\theta^{-1} \check{w}|^2_{L^2(Q)} \Big).$$

Finally, combining (4.74), (4.75) and (4.76), we obtain the desired estimate (4.57). This completes the proof of Theorem 4.3.

4.2 Exact Controllability for Semilinear Hyperbolic Equations

Fix a function $f(\cdot) \in C^1(\mathbb{R})$ satisfying the following condition:

$$\overline{\lim_{s \to \infty}} \frac{f(s)}{s \ln^{\tilde{r}} |s|} = 0, \tag{4.77}$$

where $\tilde{r} \in [0, \frac{3}{2})$. Note that the above $f(\cdot)$ may have a superlinear growth. We consider the following controlled semilinear hyperbolic equation with an internal local controller acting on ω:

$$\begin{cases} y_{tt} - \sum_{j,k=1}^{n} (h^{jk} y_{x_j})_{x_k} = \chi_\omega u + f(y) & \text{in } Q, \\ y = 0 & \text{on } \Sigma, \\ y(0) = y_0, \quad y_t(0) = y_1 & \text{in } \Omega. \end{cases} \tag{4.78}$$

In the Eq. (4.78), $(y(t, \cdot), y_t(t, \cdot))$ is the *state* and $u(t, \cdot)$ is the *control*. In what follows, we choose the *state* and the *control spaces* of the system (4.78) to be $H_0^1(\Omega) \times L^2(\Omega)$ and $L^2((0, T) \times \omega)$, respectively. For any $(y_0, y_1) \in H_0^1(\Omega) \times L^2(\Omega)$ and $u \in L^2((0, T) \times \omega)$, (4.78) admits a unique weak solution $y \in C([0, T]; H_0^1(\Omega)) \cap C^1([0, T]; L^2(\Omega))$ (e.g. [6]).

Our aim is to study the exact controllability of (4.78), by which we mean that, *for any given* $(y_0, y_1), (\tilde{y}_0, \tilde{y}_1) \in H_0^1(\Omega) \times L^2(\Omega)$, *find (if possible) a control* $u \in L^2((0, T) \times \omega)$ *such that the corresponding solution* y *to (4.78) satisfies*

$$y(T) = \tilde{y}_0, \quad y_t(T) = \tilde{y}_1 \quad \text{in } \Omega. \tag{4.79}$$

The exact controllability problem for linear and semilinear hyperbolic equations (for example, $f(\cdot)$ is a linear function, or simply, $f(\cdot) \equiv 0$ in (4.78), has been studied by many authors (e.g. [3, 10–12, 23, 28, 35, 36] and the references cited therein). We refer to [21] for local controllability results for quasilinear hyperbolic systems.

In order to obtain the exact controllability of (4.78), by the well-known duality argument (e.g., [23] and [22, p. 282, Lemma 2.4]), one needs to establish a suitable observability estimate for the following adjoint system of the linearized system of (4.78):

$$\begin{cases} z_{tt} - \sum_{j,k=1}^{n} (h^{jk} z_{x_j})_{x_k} = az & \text{in } Q, \\ z = 0 & \text{on } \Sigma, \\ z(0) = z_0 \quad z_t(0) = z_1 & \text{in } \Omega, \end{cases} \tag{4.80}$$

where $a \in L^\infty(0, T; L^p(\Omega))$ for some $p \in [n, \infty]$. More precisely, we need to derive the following observability result for (4.80):

Theorem 4.4 *Let Condition 4.2 hold, and ω and T_1 be given in (4.9) and (4.28), respectively. Then for any $T > T_1$, there is a constant $C > 0$ such that, for any $(z_0, z_1) \in L^2(\Omega) \times H^{-1}(\Omega)$, the corresponding solution z to (4.80) satisfies*

$$|(z_0, z_1)|_{L^2(\Omega) \times H^{-1}(\Omega)} \leq \mathscr{C}(r_1)|z|_{L^2((0,T) \times \omega)}, \tag{4.81}$$

where r_1 was given in (3.52) and $\mathscr{C}(r_1) \overset{\triangle}{=} \exp\left[C\left(1 + r_1^{\frac{1}{3/2-n/p}}\right)\right]$.

Proof The proof is divided into several steps.

Step 1. Since the solution z to (4.80) does not necessarily vanish at $t = 0, T$, we need to introduce a suitable cut-off function. To this end, set

$$\tilde{T}_j = \frac{T}{2} - \varepsilon_j T, \quad \tilde{T}'_j = \frac{T}{2} + \varepsilon_j T, \quad \text{and} \quad R_0 = \min_{x \in \overline{\Omega}} \sqrt{\psi(x)}, \tag{4.82}$$

where $j = 0, 1$ and $0 < \varepsilon_0 < \varepsilon_1 < \frac{1}{2}$. By (4.10), (4.28) and (4.29), for any $T > T_1$, we have that

$$\phi(0, x) = \phi(T, x) < R_1^2 - c_1 T^2 / 4 < 0, \qquad \forall\, x \in \Omega, \tag{4.83}$$

where R_1 is given by (4.13). Consequently, there exists an $\varepsilon_1 \in (0, 1/2)$, close to $1/2$, such that

$$\phi(t, x) \le R_1^2 / 2 - c_1 T^2 / 8 < 0, \qquad \forall\, (t, x) \in \left((0, \tilde{T}_1) \bigcup (\tilde{T}'_1, T) \right) \times \Omega, \tag{4.84}$$

with $\tilde{T}_1$ and $\tilde{T}'_1$ given by (4.82). Further, it follows from (4.10) that

$$\phi(T/2, x) = \psi(x) \ge R_0^2, \qquad \forall\, x \in \Omega.$$

Hence, one can find an $\varepsilon_0 \in (0, 1/2)$, close to 0, such that

$$\phi(t, x) \ge R_0^2 / 2, \qquad \forall\, (t, x) \in (\tilde{T}_0, \tilde{T}'_0) \times \Omega, \tag{4.85}$$

with $\tilde{T}_0$ and $\tilde{T}'_0$ given by (4.82). We now choose a nonnegative function $\xi \in C_0^\infty([0, T])$ such that

$$\xi(t) \equiv 1 \quad \text{in } (\tilde{T}_1, \tilde{T}'_1). \tag{4.86}$$

Clearly, ξz vanishes at $t = 0, T$. Hence, by Theorem 4.3, for any $\lambda \ge \lambda_0^*$, we have

$$\lambda \int_Q \theta^2 (\xi z)^2 dx dt$$
$$\le C \Bigg[\Bigg| \theta \Bigg\{ \Big[(\xi z)_{tt} - \sum_{j,k=1}^n \left(h^{jk} (\xi z)_{x_j} \right)_{x_k} \Big] - a \xi z \Bigg\} \Bigg|_{H^{-1}(Q)}^2 \tag{4.87}$$
$$+ \frac{1}{\lambda^{2(1-n/p)}} \left| \theta a \xi z \right|_{L^2(0,T;\, H^{-n/p}(\Omega))}^2 + \lambda^2 \int_0^T \int_\omega \theta^2 z^2 dx dt \Bigg].$$

By (4.80), noting (4.84) and (4.86), we have

$$\Bigg| \theta \Bigg\{ \Big[(\xi z)_{tt} - \sum_{j,k=1}^n \left(h^{jk} (\xi z)_{x_j} \right)_{x_k} \Big] - a \xi z \Bigg\} \Bigg|_{H^{-1}(Q)}$$
$$= \left| \theta \left(2 \xi_t z_t + z \xi_{tt} \right) \right|_{H^{-1}(Q)}$$
$$= \sup_{|f|_{H_0^1(Q)} = 1} \int_Q \left(\theta \left(2 \xi_t z_t + z \xi_{tt} \right), f \right)_{H^{-1}(Q), H_0^1(Q)} dx dt \tag{4.88}$$

$$= \sup_{|f|_{H_0^1(Q)}=1} \int_Q \theta z \left(-\xi_{tt} f - 2\xi_t f_t - 2\lambda \phi_t \xi_t f \right) dx dt$$

$$\leq C e^{(R_1^2/2 - c_1 T^2/8)\lambda}(1+\lambda)|z|_{L^2(J\times\Omega)},$$

where $J \triangleq (0, \tilde{T}_1) \cup (\tilde{T}_1', T)$.

Step 2. Recalling (3.52) for the definition of r_1 and noting that the Sobolev embedding $H_0^{n/p}(\Omega) \hookrightarrow L^{\frac{2p}{p-2}}(\Omega)$, implies (by duality) the embedding $L^{\frac{2p}{p+2}}(\Omega) \hookrightarrow H^{-n/p}(\Omega)$, we get

$$\left| e^{\lambda\phi} a\xi z \right|_{L^2(0,T; H^{-n/p}(\Omega))} \leq \left| e^{\lambda\phi} a\xi z \right|_{L^2(0,T;L^{2p/(p+2)}(\Omega))} \leq Cr_1 \left| e^{\lambda\phi} z \right|_{L^2(Q)}. \quad (4.89)$$

Further, by (4.84) and (4.86), we have

$$\int_Q e^{2\lambda\phi}(\xi z)^2 dx dt = \int_Q e^{2\lambda\phi} z^2 dx dt - \int_Q e^{2\lambda\phi}(1-\xi^2)z^2 dx dt$$

$$\geq \int_Q e^{2\lambda\phi} z^2 dx dt - C e^{(R_1^2 - cT^2/4)\lambda}|z|_{L^2(J\times\Omega)}^2. \quad (4.90)$$

Combining (4.87)–(4.90), we conclude that there is a constant $C_1 = C_1(T, \Omega)$, independent of λ and r_1, such that

$$\lambda \int_Q e^{2\lambda\phi} z^2 dx dt \leq C_1 \Big[\frac{r_1^2}{\lambda^{2(1-n/p)}} \int_Q e^{2\lambda\phi} z^2 dx dt + \lambda^2 \int_0^T \int_\omega e^{2\lambda\phi} z^2 dx dt$$

$$+ e^{(R_1^2 - c_1 T^2/4)\lambda}(1+\lambda^2)|z|_{L^2(J\times\Omega)}^2 \Big]. \quad (4.91)$$

Since $R_1^2 - c_1 T^2/4 < 0$, one may find $\lambda_1' \geq \lambda_0$ such that $e^{(R_1^2 - c_1 T^2/4)\lambda}(1+\lambda^2) < 1$ for all $\lambda \geq \lambda_1'$. Take $\lambda_1 \geq \lambda_1'$ satisfying

$$\lambda \geq r_1^{\frac{1}{3/2-n/p}} \lambda_1 \implies \lambda - C_1 \frac{r_1^2}{\lambda^{2(1-n/p)}} \geq \frac{\lambda}{2}.$$

For such a choice of λ_1, it follows from (4.91) that

$$\lambda \int_Q z^2 e^{2\lambda\phi} dx dt \leq C \Big(\lambda^2 \int_0^T \int_\omega z^2 e^{2\lambda\phi} dx dt + |z|_{L^2(J\times\Omega)}^2 \Big). \quad (4.92)$$

Step 3. Now we use a modified energy method. From (4.85), we see that

$$\int_Q e^{2\lambda\phi} z^2 dx dt \geq e^{R_0^2\lambda} \int_{\tilde{T}_0}^{\tilde{T}_0'} \int_\Omega z^2 dx dt. \quad (4.93)$$

Put

$$E(t) \triangleq \frac{1}{2} \Big(|z(t, \cdot)|_{L^2(\Omega)}^2 + |z_t(t, \cdot)|_{H^{-1}(\Omega)}^2 \Big). \quad (4.94)$$

For any $S_0 \in (\tilde{T}_0, T/2)$ and $S_0' \in (T/2, \tilde{T}_0')$, by means of the classical energy estimate, one has

$$\int_{S_0}^{S_0'} E(t)dt \le C(1 + r_1) \int_{\tilde{T}_0}^{\tilde{T}_0'} \int_{\Omega} z^2 dx dt. \tag{4.95}$$

We claim that, there is a constant $C > 0$ such that

$$E(t) \le C e^{Cr_1^{\frac{1}{2-n/p}}} E(s), \qquad \forall\, t, s \in [0, T]. \tag{4.96}$$

Note however that this does not follow from the usual energy method. Instead, we need to use the duality argument and adopt a modified energy estimate introduced in [9]. For this, for any $(w^0, w^1) \in H_0^1(\Omega) \times L^2(\Omega)$, consider the following equation:

$$\begin{cases} w_{tt} - \displaystyle\sum_{j,k=1}^{n} \left(h^{jk}(x)w_{x_j}\right)_{x_k} = aw, & \text{in } Q, \\ w = 0, & \text{on } \Sigma, \\ w(T) = w^0, \quad w_t(T) = w^1, & \text{in } \Omega. \end{cases} \tag{4.97}$$

Denote a (modified) energy of system (4.97) by

$$\mathcal{E}(t) = \frac{1}{2} \int_{\Omega} \left(|w_t|^2 + \sum_{j,k=1}^{n} h^{jk} w_{x_j} w_{x_k} + r_1^{\frac{2}{2-n/p}} |w|^2\right) dx. \tag{4.98}$$

Then, by (4.97) and the definition of r_1 in (3.52), we have

$$\frac{d\mathcal{E}(t)}{dt} = \int_{\Omega} aww_t dx + r_1^{\frac{2}{2-n/p}} \int_{\Omega} ww_t dx. \tag{4.99}$$

Put $p_1 = \frac{2p}{n-2}$ and $p_2 = \frac{2p}{p-n}$. Noting that $\frac{1}{p} + \frac{1}{p_1} + \frac{1}{p_2} + \frac{1}{2} = 1$ and $\frac{1}{2(n/p)^{-1}} + \frac{1}{2(1-n/p)^{-1}} + \frac{1}{2} = 1$, by Hölder's inequality and Sobolev's embedding theorem, and recalling (4.98), we get

$$\begin{aligned}
\int_{\Omega} aww_t dx &\le \int_{\Omega} |a||w|^{\frac{n}{p}} |w|^{1-\frac{n}{p}} |z_t| dx \\
&\le r_1 \Big| |w(t, \cdot)|^{\frac{n}{p}} \Big|_{L^{p_1}(\Omega)} \Big| |w(t, \cdot)|^{1-\frac{n}{p}} \Big|_{L^{p_2}(\Omega)} \big|w_t(t, \cdot)\big|_{L^2(\Omega)} \\
&= r_1 \big|w(t, \cdot)\big|^{\frac{n}{p}}_{L^{\frac{2n}{n-2}}(\Omega)} \big|w(t, \cdot)\big|^{1-\frac{n}{p}}_{L^2(\Omega)} \big|w_t(t, \cdot)\big|_{L^2(\Omega)} \\
&= \underbrace{r_1^{\frac{1}{2-n/p}} \big|w(t, \cdot)\big|^{\frac{n}{p}}_{L^{\frac{2n}{n-2}}(\Omega)}}_{\le \mathcal{E}(t)^{\frac{n}{2p}}} \underbrace{\left(r_1^{\frac{1-n/p}{2-n/p}} \big|w(t, \cdot)\big|^{1-\frac{n}{p}}_{L^2(\Omega)}\right)}_{\le \mathcal{E}(t)^{\frac{1}{2}-\frac{n}{2p}}} \underbrace{\big|w_t(t, \cdot)\big|_{L^2(\Omega)}}_{\le \mathcal{E}(t)^{1/2}} \\
&\le Cr_1^{\frac{1}{2-n/p}} \mathcal{E}(t).
\end{aligned} \tag{4.100}$$

Similarly,

$$r_1^{\frac{2}{2-n/p}} \int_\Omega w w_t dx \le \frac{r_1^{\frac{1}{2-n/p}}}{2} \int_\Omega \left(r_1^{\frac{2}{2-n/p}} w^2 + w_t^2 \right) dx \le C r_1^{\frac{1}{2-n/p}} \mathscr{E}(t). \qquad (4.101)$$

Combining (4.99)–(4.101), we conclude that

$$\frac{d\mathscr{E}(t)}{dt} \le C r_1^{\frac{1}{2-n/p}} \mathscr{E}(t).$$

By this and noting the time reversibility of the system (4.97), we get that

$$\mathscr{E}(t) \le C e^{C r_1^{\frac{1}{2-n/p}}} \mathscr{E}(s), \qquad \forall\, t, s \in [0, T].$$

Hence, for all $t, s \in [0, T]$,

$$|(w(t), w_t(t))|_{H_0^1(\Omega) \times L^2(\Omega)} \le C e^{C r_1^{\frac{1}{2-n/p}}} |(w(s), w_t(s))|_{H_0^1(\Omega) \times L^2(\Omega)}. \qquad (4.102)$$

Now, taking the scalar product of the first equation of (4.80) by w, integrating it in $(t, s) \times \Omega$, by (4.97) and using integration by parts, we get

$$\begin{aligned}(z(s), w_t(s))_{L^2(\Omega)} + \langle z_t(s), -w(s) \rangle_{H^{-1}(\Omega), H_0^1(\Omega)} \\ = (z(t), w_t(t))_{L^2(\Omega)} + \langle z_t(t), -w(t) \rangle_{H^{-1}(\Omega), H_0^1(\Omega)}, \qquad \forall\, t, s \in [0, T].\end{aligned} \qquad (4.103)$$

Hence, by (4.94) and (4.103), and noting the last equation in (4.97), and using (4.102), we get (denoting by S the unit sphere of the space $H_0^1(\Omega) \times L^2(\Omega)$)

$$\begin{aligned}\sqrt{2E(T)} &= \sup_{(w^0, w^1) \in S} \left[(z(T), w^1)_{L^2(\Omega)} + \langle z_t(T), -w^0 \rangle_{H^{-1}(\Omega), H_0^1(\Omega)} \right] \\ &= \sup_{(w^0, w^1) \in S} \left[(z(t), w_t(t))_{L^2(\Omega)} + \langle z_t(t), -w(t) \rangle_{H^{-1}(\Omega), H_0^1(\Omega)} \right] \\ &\le C \sqrt{E(t)} \sup_{(w^0, w^1) \in S} |(w(t), w_t(t))|_{H_0^1(\Omega) \times L^2(\Omega)} \\ &\le C e^{C r_1^{\frac{1}{2-n/p}}} \sqrt{E(t)} \sup_{(w^0, w^1) \in S} |(w(T), w_t(T))|_{H_0^1(\Omega) \times L^2(\Omega)} \\ &= C e^{C r_1^{\frac{1}{2-n/p}}} \sqrt{E(t)}.\end{aligned}$$

This fact, together with the time reversibility of the system (4.80), yields the desired estimate (4.96).

   Step 4. We now return to the proof of the observability estimate. By (4.96), we get

$$|z|^2_{L^2(J \times \Omega)} \leq C E(0) e^{C r_1^{\frac{1}{2-n/p}}}, \tag{4.104}$$

and

$$\int_{S_0}^{S_0'} E(t) dt \geq \frac{1}{C} E(0) e^{-C r_1^{\frac{1}{2-n/p}}}. \tag{4.105}$$

Combining (4.105) with (4.93) and (4.95), we get

$$\lambda \int_Q z^2 dx dt \geq \frac{\lambda}{C(1 + r_1)} e^{R_0^2 \lambda - C r_1^{\frac{1}{2-n/p}}} E(0). \tag{4.106}$$

Inequality (4.92) together with (4.104) and (4.106) yields that there is a constant $C_2 > 0$ such that for all

$$\lambda \geq \left(1 + r_1^{\frac{1}{3/2-n/p}}\right) \lambda_1 \Longrightarrow$$
$$\underbrace{\left[\lambda e^{R_0^2 \lambda - C_2 r_1^{\frac{1}{2-n/p}}} - C_2(1 + r_1) e^{C_2 r_1^{\frac{1}{2-n/p}}}\right] E(0)}_{\alpha(\lambda, r_1)} \tag{4.107}$$
$$\leq C_2 \lambda^2 (1 + r_1) e^{C_2 \lambda} \int_0^T \int_\omega z^2 dx dt.$$

Assume that $\lambda \geq \left(1 + r_1^{\frac{1}{3/2-n/p}}\right) \lambda_1$. Taking, if necessary, a greater λ_1 we have that

$$\lambda e^{\frac{R_0^2 \lambda}{2}} \geq 1 + C_2(1 + r_1), \qquad \frac{R_0^2 \lambda}{2} \geq 2 C_2 r_1^{\frac{1}{2-n/p}}. \tag{4.108}$$

(To obtain the observability inequality we used that $\frac{1}{2-n/p} < \frac{1}{3/2-n/p}$). Thus,

$$\alpha(\lambda, r_1) \geq e^{C_2 r_1^{\frac{1}{2-n/p}}} \geq 1. \tag{4.109}$$

Then, from (4.107) and (4.109), we obtain that

$$\exists \lambda_1, \quad \lambda \geq \left(1 + r_1^{\frac{1}{3/2-n/p}}\right) \lambda_1 \Longrightarrow E(0) \leq C_2 \lambda^2 (1 + r_1) e^{C_2 \lambda} |z|^2_{L^2((0,T) \times \omega)}.$$

Taking the preceding inequality at $\lambda = \left(1 + r_1^{\frac{1}{3/2-n/p}}\right) \lambda_1$ gives the desired observability inequality (4.81). This completes the proof of Theorem 4.4.

Remark 4.5 The observability estimate in the form of (4.81) was first proved in [9] (See also [12, 32] for some earlier results). In case of the hyperbolic system with N equations, it is shown in [9] that the exponent $\frac{2}{3}$ of $|a|^{\frac{2}{3}}_{L^\infty(0,T;L^\infty(\Omega;\mathbb{R}^{N \times N}))}$ in (4.81) (for the special case $p = \infty$) is sharp for $n \geq 2$ and $N \geq 2$. However, it is unknown whether the estimate is optimal for the case that $p < \infty$.

Thanks to the classical duality argument and the fixed point technique, one can show the following exact controllability result.

Theorem 4.5 *Let Condition 4.2 hold, ω and T_1 be given respectively in (4.9) and (4.28), and $T > T_1$. Then the system (4.78) is exactly controllable in $H_0^1(\Omega) \times L^2(\Omega)$ at time T.*

Proof The proof is very close to [20, Theorem 3.1] and [33, Theorem 2.1]. We divide it into two steps.

Step 1. Define a function $\eta_3(\cdot) \in C(\mathbb{R})$ by

$$\eta_3(s) \triangleq \begin{cases} [f(s) - f(0)]/s, & \text{if } s \neq 0, \\ f'(0), & \text{if } s = 0. \end{cases} \tag{4.110}$$

Let the initial and final data $(y_0, y_1), (\tilde{y}_0, \tilde{y}_1) \in H_0^1(\Omega) \times L^2(\Omega)$ be given. For any given $q(\cdot) \in L^\infty(0, T; L^2(\Omega))$, we look for a control $u = u(q(\cdot)) \in L^2((0, T) \times \omega)$ such that the solution $y = y(\cdot; q(\cdot))$ of

$$\begin{cases} y_{tt} - \sum_{j,k=1}^{n} (h^{jk} y_{x_j})_{x_k} = \eta_3(q(\cdot))y + f(0) + \chi_\omega(x)u(t, x) & \text{in } Q, \\ y = 0 & \text{on } \Sigma, \\ y(0) = y_0, \quad y_t(0) = y_1 & \text{in } \Omega \end{cases} \tag{4.111}$$

satisfies

$$y(T) = \tilde{y}_0, \quad y_t(T) = \tilde{y}_1 \quad \text{in } \Omega. \tag{4.112}$$

First, we solve the following equation:

$$\begin{cases} v_{tt} - \sum_{j,k=1}^{n} (h^{jk} v_{x_j})_{x_k} = \eta_3(q(\cdot))v + f(0) & \text{in } Q, \\ v = 0 & \text{on } \Sigma, \\ v(T) = \tilde{y}_0, \quad v_t(T) = \tilde{y}_1 & \text{in } \Omega, \end{cases} \tag{4.113}$$

which admits a unique solution $v = v(\cdot; q(\cdot)) \in C([0, T]; H_0^1(\Omega)) \cap C^1([0, T]; L^2(\Omega))$. Also, for any $(z_0, z_1) \in L^2(\Omega) \times H^{-1}(\Omega)$, we solve

$$\begin{cases} z_{tt} - \sum_{j,k=1}^{n} (h^{jk} z_{x_j})_{x_k} = \eta_3(q(\cdot))z & \text{in } Q, \\ z = 0 & \text{on } \Sigma, \\ z(0) = z_0, \quad z_t(0) = z_1 & \text{in } \Omega \end{cases} \tag{4.114}$$

and

$$\begin{cases} \eta_{tt} - \displaystyle\sum_{j,k=1}^{n} (h^{jk}\eta_{x_j})_{x_k} = \eta_3(q(\cdot))\eta + \chi_\omega(x)z(t,x) & \text{in } Q, \\ \eta = 0 & \text{on } \Sigma, \\ \eta(T) = \eta_t(T) = 0 & \text{in } \Omega. \end{cases} \quad (4.115)$$

Define a linear continuous operator $\Lambda : L^2(\Omega) \times H^{-1}(\Omega) \to H_0^1(\Omega) \times L^2(\Omega)$ by

$$\Lambda(z_0, z_1) \overset{\triangle}{=} (-\eta_t(0), \eta(0)), \quad (4.116)$$

where $\eta \in C([0, T]; H_0^1(\Omega)) \cap C^1([0, T]; L^2(\Omega))$ is the weak solution of (4.115).
Let us show the existence of some $(z_0, z_1) \in L^2(\Omega) \times H^{-1}(\Omega)$ such that

$$\Lambda(z_0, z_1) = (-y_1 + v_t(0), y_0 - v(0)). \quad (4.117)$$

By multiplying the first equation in (4.115) by w, integrating it in Q, using integration by parts, and noting (4.114), $\eta(T) = \eta_t(T) = 0$ in Ω and (4.116), we have that

$$\begin{aligned} &\int_\Omega \eta_t(0)z_0 dx - \langle \eta(0), z_1 \rangle_{H_0^1(\Omega), H^{-1}(\Omega)} \\ &= \langle \Lambda(z_0, z_1), (z_0, z_1) \rangle_{L^2(\Omega) \times H_0^1(\Omega), L^2(\Omega) \times H^{-1}(\Omega)} = \int_0^T \int_\omega z^2 dx dt. \end{aligned} \quad (4.118)$$

It follows from Theorem 4.4 and (4.118) that

$$\begin{aligned} &\langle \Lambda(z_0, z_1), (z_0, z_1) \rangle_{L^2(\Omega) \times H_0^1(\Omega), L^2(\Omega) \times H^{-1}(\Omega)} \\ &\geq \frac{1}{\mathscr{C}(r_3)} |(z_0, z_1)|^2_{L^2(\Omega) \times H^{-1}(\Omega)}, \quad \forall (z_0, z_1) \in L^2(\Omega) \times H^{-1}(\Omega), \end{aligned} \quad (4.119)$$

where $r_3 \overset{\triangle}{=} |\eta_3(q(\cdot))|_{L^\infty(0,T;L^p(\Omega))}$ for $p \in [n, \infty]$ and $\mathscr{C}(\cdot)$ is the constant given in (4.81). By Lax-Milgram theorem, the Eq. (4.117) admits a unique solution $(z_0, z_1) \in L^2(\Omega) \times H^{-1}(\Omega)$. It is easy to check that

$$u = z \quad (4.120)$$

is the desired control such that the weak solution $y \equiv v + \eta$ to (4.111) satisfies (4.112).

Step 2. We need to give an estimate on the control u given by (4.120). Concerning the system (4.113), by means of the usual energy estimate, we obtain that

$$|(v(0), v_t(0))|_{H_0^1(\Omega) \times L^2(\Omega)} \leq Ce^{Cr_3} \left(|f(0)|_{L^2(\Omega)} + |(\tilde{y}_0, \tilde{y}_1)|_{H_0^1(\Omega) \times L^2(\Omega)} \right). \quad (4.121)$$

On the other hand, by (4.116), (4.119) and (4.121), we get (recall that $r_3 \overset{\triangle}{=}$ $|\eta_3(z(\cdot))|_{L^\infty(0,T;L^p(\Omega))}$ for $p \in [n, \infty]$)

$$
\begin{aligned}
&|(z_0, z_1)|_{L^2(\Omega) \times H^{-1}(\Omega)} \\
&\leq \mathscr{C}(r_3) |\Lambda(z_0, z_1)|_{L^2(\Omega) \times H_0^1(\Omega)} \\
&= \mathscr{C}(r_3) |(-y_1 + v_t(0), y_0 - v(0))|_{L^2(\Omega) \times H_0^1(\Omega)} \\
&\leq \mathscr{C}(r_3) \Big(|(y_0, y_1)|_{H_0^1(\Omega) \times L^2(\Omega)} + |(v(0), v_t(0))|_{H_0^1(\Omega) \times L^2(\Omega)} \Big) \\
&\leq \mathscr{C}(r_3) \Big(|f(0)|_{L^2(\Omega)} + |(y_0, y_1)|_{H_0^1(\Omega) \times L^2(\Omega)} + |(\tilde{y}_0, \tilde{y}_1)|_{H_0^1(\Omega) \times L^2(\Omega)} \Big).
\end{aligned}
\tag{4.122}
$$

Using the energy method again, we have that

$$
|z|_{C([0,T];L^2(\Omega))} \leq e^{Cr_3} |(z_0, z_1)|_{L^2(\Omega) \times H^{-1}(\Omega)}.
\tag{4.123}
$$

Thus, combining (4.122) and (4.123), we obtain that

$$
\begin{aligned}
&|z|_{C([0,T];L^2(\Omega))} \\
&\leq \mathscr{C}(r_3) \Big(|f(0)|_{L^2(\Omega)} + |(y_0, y_1)|_{H_0^1(\Omega) \times L^2(\Omega)} + |(\tilde{y}_0, \tilde{y}_1)|_{H_0^1(\Omega) \times L^2(\Omega)} \Big).
\end{aligned}
\tag{4.124}
$$

Next, applying the classical energy method to (4.111) and noting (4.120)–(4.124), recalling the assumption (4.77), we conclude that there is a constant $C > 0$ such that, for any $\varepsilon \in (0, 4]$, it holds

$$
\begin{aligned}
&|y|_{C([0,T];H_0^1(\Omega)) \cap C^1([0,T];L^2(\Omega))} \\
&\leq C \Big(|f(0)| + |(y_0, y_1)|_{H_0^1(\Omega) \times L^2(\Omega)} \\
&\quad + |(\tilde{y}_0, \tilde{y}_1)|_{H_0^1(\Omega) \times L^2(\Omega)} \Big) \Big(1 + |q|_{L^\infty(0,T;L^2(\Omega))}^{4/(1+\varepsilon)} \Big).
\end{aligned}
\tag{4.125}
$$

Consequently if we take $\varepsilon = 4$ in (4.125), the desired exact controllability result follows from the fixed point technique. This completes the proof of Theorem 4.5.

Remark 4.6 Due to the blow-up and the finite propagation speed of solutions to hyperbolic equations, one cannot expect the exact controllability of the Eq. (4.78) for nonlinearities satisfying (4.77) with $\tilde{r} > 2$. One could expect the system to be exactly controllable for $\tilde{r} \leq 2$ with localized controls. However, in view of the observability estimate (4.81), the usual fixed point method cannot be applied for $\tilde{r} \geq \frac{3}{2}$. Therefore, when $n \geq 2$, the (global) exact controllability problem for the system (4.78) is still open for $\frac{3}{2} \leq \tilde{r} \leq 2$.

Remark 4.7 The assumption on the time T_1 in (4.28) plays a key role in the estimate on the boundary term (see Step 4 in the proof of Theorem 4.2). If one considers the special case, i.e. $(h^{jk})_{1 \leq j,k \leq n} = I_n$, then $s_0 = 1$ and $\psi(x) = |x - x_0|^2$, the corresponding condition on T in Theorems 4.4 and 4.5 is the following:

$$
T > T_1 = \max \Big\{ 4 \max_{x \in \overline{\Omega}} |x - x_0|, \quad 1 + 48\sqrt{n}(n+2) \max_{x \in \Gamma} \big[(x - x_0) \cdot v(x) \big] \Big\}.
$$

Hence, the restriction on T in Theorems 4.4 and 4.5 is technical, and T_1 is far from sharp. It is reasonable to expect that it can be improved to a better one as that in [3], but this is an unsolved problem.

4.3 Exponential Decay of Locally Damped Hyperbolic Equations

Based on the global Carleman estimate (4.57) in Theorem 4.3, in this section, we present an exponential decay result for hyperbolic equations.

Let $\tilde{a} \in L^\infty(\Omega)$ be a nonnegative function satisfying

$$\exists\, \tilde{a}_0 > 0 : \tilde{a}(x) \geq \tilde{a}_0, \quad \forall x \in \omega, \tag{4.126}$$

where the subset ω is a neighborhood of Γ_0 given by (4.9). Consider the following damped hyperbolic equation:

$$\begin{cases} y_{tt} - \displaystyle\sum_{j,k=1}^{n} (h^{jk} y_{x_j})_{x_k} + \tilde{a} y_t = 0 & \text{in } (0, \infty) \times \Omega, \\ y = 0 & \text{on } (0, \infty) \times \Gamma, \\ y(0) = y_0, \quad y_t(0) = y_1 & \text{in } \Omega. \end{cases} \tag{4.127}$$

For any $(y_0, y_1) \in H_0^1(\Omega) \times L^2(\Omega)$, the system (4.127) admits a unique solution $y \in C([0, \infty); H_0^1(\Omega)) \cap C^1([0, \infty); L^2(\Omega))$.

Define the energy of the system (4.127) by

$$E(t) = \frac{1}{2} \int_\Omega \Big[|y_t(t)|^2 + \sum_{j,k=1}^{n} h^{jk} y_{x_j}(t) y_{x_k}(t) \Big] dx. \tag{4.128}$$

Multiplying the Eq. (4.127) by y_t, integrating it on $(s, t) \times \Omega$ and using integration by parts, we get

$$E(t) - E(s) = -\int_\Omega \tilde{a} y_t^2 dx, \quad \forall\, 0 \leq s < t < \infty. \tag{4.129}$$

By (4.129), it is clear that the energy $E(\cdot)$ is a nonincreasing function of the time variable t. The aim of this section is to analyze the longtime behavior of $E(t)$ as the time $t \to \infty$. More precisely, we have the following result.

Theorem 4.6 *Assume that Condition 4.2 holds. Then there exist positive constants C_0 and α such that for any $(y^0, y^1) \in H_0^1(\Omega) \times L^2(\Omega)$, the energy $E(\cdot)$ of each solution to (4.127) satisfies that*

$$E(t) \leq C_0 e^{-\alpha t} E(0). \tag{4.130}$$

Proof Let us divide the proof into three steps.

Step 1. Recalling (4.82) for $\widetilde{T}_j$ and $\widetilde{T}'_j$ ($j = 0, 1$), we choose a nonnegative cut-off function $\tilde{\zeta} \in C_0^2([0, T])$ such that

$$\tilde{\zeta} \equiv 1 \text{ in } [\tilde{T}_1, \tilde{T}_1'].$$
(4.131)

Set $z(t, x) = \tilde{\zeta}(t) y_t(x, t)$ for $(t, x) \in Q$. Then, z solves

$$\begin{cases} z_{tt} - \sum_{j,k=1}^{n} (h^{jk} z_{x_j})_{x_k} = \tilde{\zeta}_{tt} y_t + 2\tilde{\zeta}_t y_{tt} - \tilde{\zeta}\tilde{a} y_{tt} & \text{in } (0, \infty) \times \Omega, \\ z = 0 & \text{on } (0, \infty) \times \Gamma, \\ z(0) = z(T) = 0 & \text{in } \Omega. \end{cases}$$
(4.132)

Let T_1 and ϕ be given by (4.28) and (4.10), respectively. Then, by Theorem 4.3 (taking $a = 0$), there exists $\lambda_0^* > 0$ such that for all $T > T_1$ and $\lambda \geq \lambda_0^*$, it holds that

$$\lambda \int_Q \theta^2 z^2 dx dt$$
$$\leq C \left(\left| \theta \left(\tilde{\zeta}_{tt} y_t + 2\tilde{\zeta}_t y_{tt} - \tilde{\zeta}\tilde{a} y_{tt} \right) \right|_{H^{-1}(Q)}^2 + \lambda^2 \int_0^T \int_\omega \theta^2 z^2 dx dt \right).$$
(4.133)

Now using Hölder's inequality and Sobolev embedding theorem, we find that

$$\begin{cases} \left| \theta \tilde{\zeta}_{tt} y_t \right|_{H^{-1}(Q)} \leq C \left| \theta y_t \right|_{L^2(\tilde{Q})}, \\ \left| 2\theta \tilde{\zeta}_t y_{tt} \right|_{H^{-1}(Q)} \leq C(1 + \lambda) \left| \theta y_t \right|_{L^2(\tilde{Q})}, \\ \left| \theta \tilde{\zeta}\tilde{a} y_{tt} \right|_{H^{-1}(Q)} \leq C(1 + \lambda) \left| \theta \tilde{a} y_t \right|_{L^2(Q)}, \end{cases}$$
(4.134)

where $\tilde{Q} = ((0, \tilde{T}_1) \cup (\tilde{T}_1', T)) \times \Omega$.

Combining (4.132)–(4.134), by (4.126), and noting that $w = \tilde{\zeta} y_t$, we have that

$$\lambda \left| \theta z \right|_{L^2(Q)}^2 \leq C(1 + \lambda) \left(\left| \theta y_t \right|_{L^2(\tilde{Q})}^2 + \left| \theta \tilde{a} y_t \right|_{L^2(Q)}^2 \right) + C\lambda^2 \left| \theta z \right|_{L^2(0,T;L^2(\omega))}^2$$
$$\leq C(1 + \lambda) \left| \theta y_t \right|_{L^2(\tilde{Q})}^2 + C(1 + \lambda^2) \int_0^T \int_\Omega \theta^2 \tilde{a} y_t^2 dx dt.$$
(4.135)

On the other hand, by (4.131), we find that

$$\left| \theta z \right|_{L^2(Q)}^2 \geq \int_{\tilde{T}_1}^{\tilde{T}_1'} \int_\Omega \theta^2 y_t^2 dx dt.$$

Thus,

$$\left| \theta y_t \right|_{L^2(Q)}^2 \leq \left| \theta z \right|_{L^2(Q)}^2 + \left| \theta y_t \right|_{L^2(\tilde{Q})}^2.$$
(4.136)

It follows from (4.135) and (4.136) that for any $\lambda \geq \lambda_0^* > 1$,

$$\lambda \left| \theta y_t \right|_{L^2(Q)}^2 \leq C\lambda \left| \theta y_t \right|_{L^2(\tilde{Q})}^2 + C\lambda^2 \int_0^T \int_\Omega \theta^2 \tilde{a} y_t^2 dx dt.$$
(4.137)

Step 2. Recalling (4.84) and (4.85) for the properties of the function ϕ, i.e. the weight function ϕ satisfying

$$\begin{cases} \phi(t, x) \le R_1^2/2 - c_1 T^2/8 < 0, & \forall (t, x) \in \widetilde{Q}, \\ \phi(t, x) \ge R_0^2/2, & \forall (t, x) \in Q_0 \overset{\triangle}{=} (\widetilde{T}_0, \widetilde{T}_0') \times \Omega, \end{cases} \tag{4.138}$$

where

$$\widetilde{T}_j = \frac{T}{2} - \varepsilon_j T, \ \ \widetilde{T}_j' = \frac{T}{2} + \varepsilon_j T, \ \ R_0 = \min_{x \in \overline{\Omega}} \sqrt{\psi(x)}, \ \ R_1 = \max_{x \in \overline{\Omega}} \sqrt{\psi(x)}, \ \ j = 1, 2,$$

and $0 < \varepsilon_0 < \varepsilon_1 < \frac{1}{2}$. Therefore, by (4.137) and (4.138), we obtain that

$$e^{\lambda R_0^2} |y_t|_{L^2(Q_0)}^2 \le C e^{\lambda(R_1^2 - c_1 T^2/4)} |y_t|_{L^2(\widetilde{Q})}^2 + C\lambda e^{2\lambda R_1^2} \int_0^T \int_\Omega \tilde{a} y_t^2 dx dt.$$

Next, by (4.129), noting that the energy is nonincreasing, and $\widetilde{Q} = ((0, \widetilde{T}_1) \cup (\widetilde{T}_1', T)) \times \Omega$, we have

$$|y_t|_{L^2(\widetilde{Q})}^2 \le 2T E(0) = 2T E(T) + 2T \int_0^T \int_\Omega \tilde{a} y_t^2 dx dt.$$

Hence,

$$|y_t|_{L^2(Q_0)}^2 \le C e^{\lambda(R_1^2 - R_0^2 - c_1 T^2/4)} E(T) + C\lambda e^{2\lambda R_1^2} \int_0^T \int_\Omega \tilde{a} y_t^2 dx dt. \tag{4.139}$$

Further, we take $\hat{\zeta} \in C^1([T_0, T_0'])$ with $\hat{\zeta}(\widetilde{T}_0) = \hat{\zeta}(\widetilde{T}_0') = 0$. Multiplying the first equation in (4.127) by $\hat{\zeta} y$, integrating it on Q_0 and using integration by parts, we get that

$$2 \int_{T_0}^{T_0'} \hat{\zeta}(t) E(t) dt = 2 \int_{Q_0} \hat{\zeta} y_t^2 dx dt + \int_{Q_0} \hat{\zeta}_t y_t y dx dt - \int_{Q_0} \hat{\zeta} \tilde{a} y_t y dx dt.$$

Thus, a simple calculation shows that

$$\int_{\widetilde{T}_0}^{\widetilde{T}_0'} \zeta(t) E(t) dt \le C \int_{Q_0} y_t^2 dx dt + C \int_{Q_0} \tilde{a} y_t^2 dx dt.$$

Then

$$E(T) \le C \int_{Q_0} y_t^2 dx dt + C \int_{Q_0} \tilde{a} y_t^2 dx dt. \tag{4.140}$$

It follows from (4.139) and (4.140) that

$$E(T) \leq Ce^{\lambda(R_1^2 - R_0^2 - c_1 T^2/4)} E(T) + C\lambda e^{2\lambda R_1^2} \int_0^T \int_\Omega \tilde{a} y_t^2 dx dt. \qquad (4.141)$$

Step 3. Let λ be large enough such that

$$Ce^{\lambda(R_1^2 - R_0^2 - c_1 T^2/4)} \leq 1/2.$$

In this case, (4.141) becomes

$$E(T) \leq C\lambda e^{2\lambda R_1^2} \int_0^T \int_\Omega \tilde{a} y_t^2 dx dt,$$

from which one can easily deduce that

$$E(T) \leq C_1 e^{C_2} \int_0^T \int_\Omega \tilde{a} y_t^2 dx dt, \qquad (4.142)$$

where C_1 and C_2 are positive constants, which are independent of the initial data. By (4.129) and (4.142), we get

$$E(T) \leq \gamma E(0),$$

where

$$\gamma = \frac{C_1 e^{C_2}}{1 + C_1 e^{C_2}} \in (0, 1).$$

Now, we use the semigroup property. Since problem (4.127) is globally well-posed, we may write the solution as $S(t)\varphi_0 = (y, y_t)^T$, where $\varphi_0 = (y_0, y_1)^T$. Consequently, $S(t + r) = S(t)S(r)$ for all $t, r \geq 0$. It follows that

$$E(nT) = E(S(nT)\varphi_0) = E(S^n(T)\varphi_0) \leq \gamma^n E(S(T)\varphi_0) \leq \gamma^n E(0),$$

which implies that

$$E(t) \leq \frac{1}{\gamma} e^{(-t \ln(1/\gamma))/T} E(0).$$

This completes the proof of Theorem 4.6.

Remark 4.8 Actually, Theorem 4.6 can be proved more directly and easily. Here, we adopt a method introduced in [29] because this method can be used to handle a more general damped hyperbolic equation. More precisely, consider the following damped hyperbolic equation:

$$\begin{cases} y_{tt} - \sum_{j,k=1}^{n} (h^{jk} y_{x_j})_{x_k} + p(x)y + f(y) + \tilde{a}g(y_t, \nabla y) = 0 & \text{in } (0, \infty) \times \Omega \\ y = 0 & \text{on } (0, \infty) \times \Gamma \\ y(0) = y^0, \quad y_t(0) = y^1 & \text{in } \Omega, \end{cases}$$

where $p(\cdot) \in L^m(\Omega)$ is nonnegative ($m = 2$ for $n = 1$, $m > 2$ for $n = 2$, and $m \geq n$ for $n \geq 3$), $f : \mathbb{R} \to \mathbb{R}$ is a differentiable function and $g : \mathbb{R}^{1+n} \to \mathbb{R}$ is a globally Lipschitz function. Under suitable assumptions on f and g, the exponential decay for the above damped hyperbolic equation was proved in [29].

4.4 Inverse Hyperbolic Problems

Consider the following equation:

$$\begin{cases} y_{tt} - \sum_{j,k=1}^{n} (h^{jk} y_{x_j})_{x_k} = q_1 \cdot \nabla y + q_2 y + Rf & \text{in } Q, \\ y = 0 & \text{on } \Sigma, \\ y(0, x) = 0, \quad y_t(0, x) = 0 & \text{in } \Omega. \end{cases} \tag{4.143}$$

Here $q_1 \in L^\infty(\Omega; \mathbb{R}^n)$, $q_2 \in L^\infty(\Omega)$ and $R \in H^1(0, T; L^\infty(\Omega))$ are known functions, while $f \in L^2(\Omega)$ is unknown. In this section, we are concerned with the following problem:

Problem (IPH): Let Γ_0 be a suitable subset of Γ and $T > 0$ be large enough. Determine $f(\cdot)$ through the boundary observation $\left. \frac{\partial y}{\partial \nu} \right|_{(0,T) \times \Gamma_0}$.

The answer to Problem (IPH) is stated as follows.

Theorem 4.7 *Assume that Condition 4.2 holds, and for some $\beta_0 > 0$,*

$$\inf_{x \in \Omega} |R(0, x)| \geq \beta_0. \tag{4.144}$$

Then, for Γ_0 given by (4.8) and any $T > \frac{T_0}{2}$ with T_0 defined in (4.13), there exists a constant $C = C(T, \Omega, \Gamma_0, q_1, q_2, \beta_0) > 0$ such that for each $f \in L^2(\Omega)$, the corresponding solution y to (4.143) satisfies

$$C^{-1} |f|_{L^2(\Omega)} \leq \left| \frac{\partial y}{\partial \nu} \right|_{H^1(0,T;L^2(\Gamma_0))} \leq C |f|_{L^2(\Omega)}. \tag{4.145}$$

Proof Let $z = y_t$. Noting that $z(0) = y_t(0) = y_1 = 0$, we see that z solves

$$\begin{cases} z_{tt} - \sum_{j,k=1}^{n} (h^{jk} z_{x_j})_{x_k} = q_1 \cdot \nabla z + q_2 z + R_t f & \text{in } Q, \\ z = 0 & \text{on } \Sigma, \\ z(0,x) = 0, \quad z_t(0,x) = f R(0,x) & \text{in } \Omega. \end{cases} \quad (4.146)$$

Let

$$\tilde{z}(t,x) = \begin{cases} z(t-T,x), & \text{if } (t,x) \in [T, 2T] \times \Omega, \\ z(T-t,x), & \text{if } (t,x) \in [0, T) \times \Omega, \end{cases}$$

and

$$\widetilde{R}(t,x) = \begin{cases} R(t-T,x), & \text{if } (t,x) \in [T, 2T] \times \Omega, \\ -R(T-t,x), & \text{if } (t,x) \in [0, T) \times \Omega. \end{cases}$$

Then, $\tilde{z} \in C([0, 2T], H_0^1(\Omega)) \cap C^1([0, 2T], L^2(\Omega))$ and $\widetilde{R} \in H^1(0, 2T; L^\infty(\Omega))$ satisfy that

$$\begin{cases} \tilde{z}_{tt} - \sum_{j,k=1}^{n} (h^{jk} \tilde{z}_{x_j})_{x_k} = q_1 \cdot \nabla \tilde{z} + q_2 \tilde{z} + \widetilde{R}_t f & \text{in } (0, 2T) \times \Omega, \\ \tilde{z} = 0 & \text{on } (0, 2T) \times \Gamma. \end{cases} \quad (4.147)$$

Recall (4.10) for the definitions of $\theta(t,x)$, $\ell(t,x)$, $\phi(t,x)$ and $\Psi(x)$. Set

$$v = \theta \tilde{z}, \quad P_1 v = v_{tt} - \Delta v + (|\ell_t|^2 - |\nabla \ell|^2) v.$$

Then,

$$2 \int_0^T \int_\Omega P_1 v v_t \, dx dt$$

$$= \int_\Omega |v_t(T)|^2 dx - \int_\Omega (|v_t(0)|^2 + |\nabla v(0)|^2) dx - \int_0^T \int_\Omega (|\ell_t|^2 - |\nabla \ell|^2)_t |v|^2 dx dt$$

$$= \int_\Omega |\theta(T) z_t(0)|^2 dx - \int_\Omega (|v_t(0)|^2 + |\nabla v(0)|^2) dx$$

$$\quad - \lambda^2 \int_0^T \int_\Omega [4c_1^2(t-T)^2 - |\nabla \psi|^2]_t |v|^2 dx dt$$

$$= \int_\Omega e^{2\ell(T)} |f R(0)|^2 dx - \int_\Omega (|v_t(0)|^2 + |\nabla v(0)|^2) dx - 8\lambda^2 \int_0^T \int_\Omega c_1^2(t-T)|v|^2 dx dt.$$

Next, by Cauchy-Schwarz inequality and Theorem 4.1, we have

$$\int_\Omega e^{2\ell(T)} |f R(0)|^2 \, dx$$

$$= 2 \int_0^T \int_\Omega P_1 v v_t \, dx dt + 8\lambda^2 \int_0^T \int_\Omega c_1^2(t-T) v^2 dx dt + \int_\Omega (|v_t(0)|^2 + |\nabla v(0)|^2) dx$$

$$
\leq \frac{2}{\sqrt{\lambda}} \left(\int_0^T \int_\Omega |P_1 v|^2 \, dx dt \right)^{\frac{1}{2}} \left(\lambda \int_0^T \int_\Omega |v_t|^2 dx dt \right)^{\frac{1}{2}} + \int_\Omega \left(|v_t(0)|^2 + |\nabla v(0)|^2 \right) dx
$$

$$
\leq \frac{C}{\sqrt{\lambda}} \left(\int_0^{2T} \int_\Omega e^{2\ell} |f R_t|^2 dx dt + \lambda^{\frac{3}{2}} \int_0^{2T} \int_{\Gamma_0} e^{2\ell} \left| \frac{\partial \tilde{z}}{\partial \nu} \right|^2 d\Gamma dt \right) \tag{4.148}
$$

$$
\leq \frac{C |R_t|_{L^2(0,2T;L^\infty(\Omega))}}{\sqrt{\lambda}} \int_\Omega e^{2\ell(T)} |f|^2 dx + C\lambda \int_0^{2T} \int_{\Gamma_0} e^{2\ell} \left| \frac{\partial \tilde{z}}{\partial \nu} \right|^2 d\Gamma dt.
$$

By taking λ large enough in (4.148), we get that

$$
\int_\Omega e^{2\ell(T)} |f|^2 dx \leq C\sqrt{\lambda} \int_0^{2T} \int_{\Gamma_0} e^{2\ell} \left| \frac{\partial \tilde{z}}{\partial \nu} \right|^2 d\Gamma dt \leq C\sqrt{\lambda} \int_0^{2T} \int_{\Gamma_0} e^{2\ell} \left| \frac{\partial y_t}{\partial \nu} \right|^2 d\Gamma dt.
$$

This completes the proof of Theorem 4.7.

4.5 Further Comments

In the end of this chapter, we list some further comments as follows:

- As in the previous two chapters, we only consider L^2-Carleman estimates in this chapter. In order to derive sharp unique continuation result for hyperbolic equations with rough potentials, people need to establish some L^p-Carleman estimates ($p > 1$) for these equations (e.g. [8, 16, 18]).
- Different from Chaps. 2 and 3, we do not study SUCP for hyperbolic equations. One may combine the Carleman estimate for elliptic operators in Sect. 2.4 and some transmutation method to prove some strong unique continuation properties for hyperbolic equations [19, 30]. We omit the details for two reasons. One is the limitation of length. Another is that the main idea and method are similarly to the proof of Theorem 2.4.
- The study of the long time behavior of solutions to hyperbolic equations with locally distributed damping has a long history. We refer the readers to [23, 28] for early works on this topic, to [2, 7, 29] for further research and [1] for a survey of recent works.
- Carleman estimates have been used to solve inverse coefficient problems of hyperbolic equations for a long time (See [5] for an early work and [17] for a survey on this topic). In this book, the measurement is performed on a suitable subset Γ_0 of the boundary, which satisfies the geometric condition (4.8). When the measurement is only possible on a part of the boundary without the above condition, by adopting the Carleman estimate in [15] and the Fourier–Bros–Iagolnitzer transform, uniqueness and logarithmic stability for an inverse coefficient problem of the wave equation were proved in [4].
- Recently, there are some works addressing the Carleman estimate for stochastic hyperbolic equations to study unique continuation and state observation problems ([13, 25, 26, 34]) and inverse problems [27, 31]. There are lots of interesting and difficult problems in this topic.

References

1. Alabau-Boussouira, F.: On some recent advances on stabilization for hyperbolic equations. In: Control of Partial Differential Equations. Lecture Notes in Mathematics, vol. 2048, pp. 1–100. Fond. CIME/CIME Found. Subser. Springer, Heidelberg (2012)
2. Alabau-Boussouira, F., Ammari, K.: Sharp energy estimates for nonlinearly locally damped PDEs via observability for the associated undamped system. J. Funct. Anal. **260**, 2424–2450 (2011)
3. Bardos, C., Lebeau, G., Rauch, J.: Sharp sufficient conditions for the observation, control and stabilization of waves from the boundary. SIAM J. Control Optim. **30**, 1024–1065 (1992)
4. Bellassoued, M., Yamamoto, M.: Logarithmic stability in determination of a coefficient in an acoustic equation by arbitrary boundary observation. J. Math. Pures Appl. **85**, 193–224 (2006)
5. Bukhgeim, A.L., Klibanov, M.V.: Global uniqueness of class of multidimensional inverse problems. Soviet Math. Dokl. **24**, 244–247 (1981)
6. Cazenave, T., Haraux, A.: Equations d'évolution avec non-linéarité logarithmique. Ann. Fac. Sci. Toulouse. **2**, 21–51 (1980)
7. Dehman, B., Lebeau, G., Zuazua, E.: Stabilization and control for the subcritical semilinear wave equation. Ann. Sci. École Norm. Sup. **36**, 525–551 (2003)
8. Dos Santos Ferreira, D.: Sharp L^p Carleman estimates and unique continuation. Duke Math. J. **129**, 503–550 (2005)
9. Duyckaerts, T., Zhang, X., Zuazua, E.: On the optimality of the observability inequalities for parabolic and hyperbolic systems with potentials. Ann. Inst. H. Poincaré Anal. Non Linéaire. **25**, 1–41 (2008)
10. Fattorini, H.O.: Local controllability of a nonlinear wave equations. Math. Syst. Theory **9**, 30–45 (1975)
11. Fattorini, H.O., Russell, D.L.: Exact controllability theorems for linear parabolic equations in one space dimension. Arch. Rational Mech. Anal. **43**, 272–292 (1971)
12. Fu, X., Yong, J., Zhang, X.: Exact controllability for the multidimensional semilinear hyperbolic equations. SIAM J. Control Optim. **46**, 1578–1614 (2007)
13. Fu, X., Liu, X., Lü, Q., Zhang, X.: An internal observability estimate for stochastic hyperbolic equations. ESAIM Control Optim. Calc. Var. **22**, 1382–1411 (2016)
14. Imanuvilov, OYu.: On Carlerman estimates for hyperbolic equations. Asymptot. Anal. **32**, 185–220 (2002)
15. Imanuvilov, OYu., Yamamoto, M.: Global Lipschitz stability in an inverse hyperbolic problem by interior observations. Inverse Probl. **17**, 717–728 (2001)
16. Kenig, C.E., Ruiz, A., Sogge, C.D.: Uniform Sobolev inequalities and unique continuation for second order constant coefficient differential operators. Duke Math. J. **55**, 329–347 (1987)
17. Klibanov, M.V.: Carleman estimates for global uniqueness, stability and numerical methods for coefficient inverse problems. J. Inverse Ill-Posed Probl. **21**, 477–560 (2013)
18. Koch, H., Tataru, D.: Dispersive estimates for principally normal pseudodifferential operators. Commun. Pure Appl. Math. **58**, 217–284 (2005)
19. Lebeau, G.: Un probléme d'unicité forte pour l'équation des ondes. Commun. Partial Differ. Equ. **24**, 777–783 (1999)
20. Li, W., Zhang, X.: Controllability of parabolic and hyperbolic equations: toward a unified theory. In: Control Theory of Partial Differential Equations. Lecture Notes in Pure and Applied Mathematics, vol. 242, pp. 157–174. Chapan & Hall/CRC, Boca Raton (2005)
21. Li, T.T.: Controllability and Observability for Quasilinear Hyperbolic Systems. AIMS Series on Applied Mathematics. vol. 3, American Institute of Mathematical Sciences (AIMS). Springfield (2010)
22. Li, X., Yong, J.: Optimal Control Theory for Infinite-Dimensional Systems. Birkhäuser Boston Inc, Boston (1995)
23. Lions, J.L.: Contrôlabilité Exacte, Perturbations et Stabilisation de Systèmes distribués, Tome 1, Contrôlabilité exacte. Recherches en Mathématiques Appliquées. vol. 8, Masson, Paris (1988)

24. Liu, Y.: Some sufficient conditions for the controllability of wave equations with variable coefficients. Acta Appl. Math. **128**, 181–191 (2013)
25. Lü, Q., Yin, Z.: Unique continuation for stochastic hyperbolic equations. arXiv:1701.03599
26. Lü, Q.: Some results on the controllability of forward stochastic heat equations with control on the drift. J. Funct. Anal. **260**, 832–851 (2011)
27. Lü, Q., Zhang, X.: Global uniqueness for an inverse stochastic hyperbolic problem with three unknowns. Commun. Pure Appl. Math. **68**, 948–963 (2015)
28. Russell, D.L.: Controllability and stabilizability theory for linear partial differential equations: recent progress and open problems. SIAM Rev. **20**, 639–739 (1978)
29. Tebou, L.: A Carleman estimate based approach for the stabilization of some locally damped semilinear hyperbolic equations. ESAIM: Control Optim. Calc. Var. **14**, 561–574 (2008)
30. Vessella, S.: Quantitative estimates of strong unique continuation for wave equations. Math. Ann. **367**, 135–164 (2017)
31. Yuan, G.: Determination of two kinds of sources simultaneously for a stochastic wave equation. Inverse Probl. **31**, 085003 (2015)
32. Zhang, X.: Explicit observability estimate for the wave equation with potential and its application. R. Soc. Lond. Proc. Ser. A Math. Phys. Eng. Sci. **456**, 1101–1115 (2000)
33. Zhang, X.: Explicit observability inequalities for the wave equation with lower order terms by means of Carleman inequalities. SIAM J. Control Optim. **39**, 812–834 (2001)
34. Zhang, X.: Carleman and observability estimates for stochastic wave equations. SIAM J. Math. Anal. **40**, 851–868 (2008)
35. Zuazua, E.: Exact controllability for semilinear wave equations in one space dimension. Ann. Inst. H. Poincaré Anal. Non Linéaire. **10**, 109–129 (1993)
36. Zuazua, E.: Exact controllability for the semilinear wave equation. J. Math. Pures Appl. **69**, 1–31 (1990)

Printed in the United States
By Bookmasters